LANGUAGE DEVELOPMENT THROUGH CONTENT

Mathematics Book A

Learning Strategies for Problem Solving

ANNA UHL CHAMOT

J. MICHAEL O'MALLEY

Addison-Wesley Publishing Company

Reading, Massachusetts • Menlo Park, California • New York • Don Mills, Ontario • Wokingham, England
Amsterdam • Bonn • Sydney • Singapore • Tokyo • Madrid • Bogota • Santiago • San Juan

A Publication of the World Language Division

Acknowledgments

Editorial Development: Terry Maxwell, Bernice Randall, Judith Bittinger
Production/Manufacturing: James W. Gibbons
Design: Design Office, San Francisco
Illustrations: Diana Thewlis, Sally Shimitzu, Phyllis Rockne

ISBN 0-201-12931-0
  5 6 7 8 9 10 11 12—AL—95 94 93 92

To the Teacher

Language Development Through Content: Mathematics A is specifically designed to help prepare limited English proficient students to work successfully in the language-related areas of mathematics. Through language and mathematics exercises, it helps students learn the language of mathematics and develop their problem-solving abilities. Most importantly, students are taught to use learning strategies derived from a cognitive model of learning that they can apply to both language and mathematics tasks. These strategies are included to help students in problem solving, in learning and remembering important concepts, and in becoming more effective and autonomous learners.

The content core of this student worktext is an overview of functional mathematics concepts and operations. These include number properties, word problems (with the four operations), fractions, measurement, estimation, graphing, and geometry. The worktext consists of transitional and preparatory material for mainstream classes; it is not intended as a substitute for mathematics textbooks. The worktext is intended for students who have developed some basic computation skills but who are encountering difficulty in solving word problems and in understanding explanations of mathematical concepts in English. After completing this worktext, students will have acquired valuable language experience in an academic context and will have developed basic strategies for solving word problems.

Language Development Through Content: Mathematics A provides an overview of functional mathematics, together with learning strategies for problem solving and an instructional approach for developing English language skills in an academic context. Both oral and written academic language skills are practiced. Students use oral language to talk about math concepts and to "think aloud" about how they solve problems. Through the discussion of problem-solving strategies with other students, they become aware of alternative ways to obtain an answer. Students practice listening to mathematical equations and oral word problems, and taking notes by transforming the information into mathematical symbols. Moreover, they practice both oral and silent reading of directions, math concepts, and word problems. Students learn to use reading strategies such as: inferencing to guess at word meanings from context, identifying the question and important words in problems, selecting the necessary data for solving a problem, recalling factual information, making interpretations, and gaining information from graphs and charts. Writing skills are developed through

planning, writing, and revising word problems that can be understood and solved by classmates. Throughout the book students practice evaluating their own work and identifying the nature of any difficulties they encounter.

Language Development Through Content: Mathematics A can be used effectively in both the ESOL classroom and the mainstream classroom. Instructions for the student activities are clear and complete. You will wish to go over them orally with the students, however, to make sure that each student understands what he or she is to do. In addition, the Teacher's Guide presents detailed suggestions for teaching each lesson. By following these suggestions, you will be providing your students with many opportunities to use their developiing English skills. You will also be helping them to build a valuable inventory of learning strategies which should carry over directly into their mainstream work. Cooperative learning activities are included in almost every lesson. For these pair and group activities, mixed (heterogeneous) grouping is recommended wherever possible, so that more proficient students can provide assistance and serve as models for those who are less proficient. Some students may be more proficient in mathematics skills but less proficient in English, while others may have less mathematics proficiency but more English proficiency. An ideal cooperative team includes both types of students.

The unit openers are intended to be used as oral exercises to provide practice in listening to numbers, equations, and word problems. **The texts for these and other listening exercises are found only in the Teacher's Guide.** The Guide also suggests how to teach test-taking strategies for the Checkup in each unit. A reproducible checklist for charting individual student progress and correct answers for all exercises in the worktext are included, as well, in the Teacher's Guide.

Language Development Through Content: Mathematics A teaches students the language, problem-solving skills, and learning strategies that will help them study mathematics in the mainstream curriculum. At the same time, it helps these students learn or review basic mathematics concepts presented at the elementary school level and in functional mathematics courses for older students. It can help your students toward greater success in their mainstream classrooms.

Contents

Number And Place Value

Look at the picture. Your teacher will tell you what to do.

Vocabulary for Math Directions

A. Your teacher will read some math directions. Listen and follow the directions.

1. _____ , _____

2. 2
     _3_

3. 11 12 13 14 15 16 17 18 19

4.

5. 25
     _13_

6. 30 24 25 29 21 28 20 26 22 27 23

____, ____, ____, ____, ____, ____, ____, ____, ____, ____, ____

7.

B. Work with a friend. Read the following directions. Write the number of the problem in Section A that each direction is for.

1. _____ Subtract the smaller number from the larger number.

2. _____ Write the numbers **fifteen** and **fifty.**

3. _____ Write these numbers in order.

4. _____ Cross out the numbers **twelve, fifteen, seventeen, nineteen.**

5. _____ Ring the two shapes that are the same.

6. _____ Add the numbers.

7. _____ Write the names of the coins.

Place Value

We use the base-ten place value system. In this system we use only 10 digits to write all numbers. These 10 digits are:
0 1 2 3 4 5 6 7 8 9.
    The **place** of a digit in a number tells us its **value,** or how much it is worth.

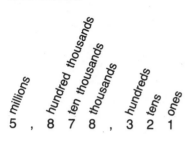

When you read this number aloud, you say: "Five million, eight hundred seventy-eight thousand, three hundred twenty-one."
    When you write a large number, use **commas** (not periods) between hundreds and thousands. For example: **5,350.** We read this as: "five thousand three hundred fifty." You also use a comma to separate hundred thousands from millions. For example: 1,832,456. We read this as: "one million, eight hundred thirty-two thousand, four hundred fifty-six."

A. Listen to your teacher read some large numbers. Write the numbers on the left. After writing all the numbers your teacher says, answer the question next to each number.

1. _____ How many ones in this number? _____

2. _____ How many hundreds in this number? _____

3. _____ How many hundred thousands in this number? _____

4. _____ How many tens in this number? _____

5. _____ How many ten thousands in this number? _____

6. _____ How many thousands in this number? _____

7. _____ How many millions in this number? _____

B. Your teacher will read two numbers. Write them on these lines. Ring the number that is **greater** in each pair.

1. _____ _____   **2.** _____ _____

3. _____ _____   **4.** _____ _____

5. _____ _____   **6.** _____ _____

C. Sometimes there are different ways to say a number. For example, for **1,500,** you can say: "one thousand five hundred." You can also say: "fifteen hundred."

    Work with a friend. Say each number two different ways.

1. 1,300   **2.** 4,800   **3.** 1,200

4, 6,700   **5.** 1,100   **6.** 9,500

D. Read each pair of numbers below. Ring the number that is **less** in each pair.

1. 7,001   5,987   **2.** 35,812   36,985   **3.** 100,000 99,999

4. 3,201,750 2,875,000   **5.** 12,856,900 15,967,088   **6.** 82,799   114,305

Check your work with a friend. Look at Section B and this section. Compare your answers. When you don't agree, ask questions for help.

E. Now play a game with your friend. You will each need a sheet of paper. Read the first group of numbers aloud. Your friend will write the five numbers without looking at his or her book.

    Then your friend will read the second group of numbers to you. You must write these five numbers without looking at your book.

1. 353   **1.** 489

2. 27   **2.** 50

3. 1,666   **3.** 2,361

4. 152,815   **4.** 161,716

5. 10,408   **5.** 12,207

Check your work. Did you remember to use a comma to separate hundreds from thousands? Write your score and your friend's score below.

My Score: _____   My Friend's Score: _____

Odd and Even Numbers

A. Write the numbers. Then ring each **even** number.

1. __1__ □ (__2__) ⊟ __3__ ⊞ (__4__) ⊞ _____ ⊞ _____ ⊞

2. _____ 〇〇〇/〇〇〇〇   _____ 〇〇〇〇/〇〇〇〇   _____ 〇〇〇〇/〇〇〇〇〇   _____ 〇〇〇〇〇/〇〇〇〇〇

3. _____ ****/****   _____ ****/*****   _____ *****/*****   _____ *****/*****

4. _____ ⊞ _____ ⊞ _____ ⊞ _____ ⊞

B. Complete the table.

Odd numbers	1								
Even numbers	2								

C. Write 10 odd numbers in Box 1.   Write 10 even numbers in Box 2.

Box 1
Odd Numbers

Box 2
Even Numbers

Work with a friend. Say one of the numbers you wrote in Box 1
or Box 2. Your friend has to say "odd" or "even." Continue to
say your numbers until your friend has had 10 turns. Write your
friend's score. Now it is your turn! Your friend will say 10
numbers. You say whether each number is odd or even. Then
write your score.

My Friend's Score: _____   My Score: _____

Skip Counting

A. Listen to your teacher count by **twos.** Write the numbers. Some numbers are already there.

<u> 2 </u> __ __ __ __ __ __ __ __ <u> 20 </u> __ __ __ __ __ __ __

B. Listen to your teacher count by **fives.** Write the numbers. Some numbers are already there.

<u> 5 </u> __ __ __ __ __ __ __ <u> 45 </u> __ __ __ __ __ __ __

C. Listen to your teacher count by **tens.** Write the numbers. Some numbers are already there.

<u> 10 </u> __ __ __ __ __ __ __ <u> 100 </u> __ __ __ __ __ __

D. Sit with one or more friends. Check your work together.

Rounding

Rounding numbers means that you say about how many or about how much.

E. Look at the heights of these mountains. Round each height to the nearest thousand feet. Write **about how high** each mountain is.

Everest	29,028 ft	About _____ ft	Demavend	18,606 ft	About _____ ft
Aconcagua	22,834 ft	About _____ ft	Ararat	16,945 ft	About _____ ft
Kilimanjaro	19,321 ft	About _____ ft			

F. Compare mountain heights. Write questions for each pair of mountains. Write complete sentences. The first one is done for you.

1. Everest — Ararat

   <u>Which mountain is higher, Everest or Ararat?</u>

2. Demavend — Kilimanjaro

   _____

3. Ararat — Aconcagua

   _____

G. Sit with one or more friends. Take turns asking and answering the questions you wrote.

Estimation and Measurement

A. Look at the chart. You are going to estimate the measurements of some things at school. **Estimate** means to guess a number that is close to the correct number.

    Follow these steps:

1. Look carefully at each picture. Estimate the measurement. Write your estimate in Column A.

2. Meet with three or four friends. Talk about your estimates. Decide on the best estimate for each thing. Write your group's estimate in Column B.

3. Use a ruler or tape measure. Measure the objects named on the chart. Do the measuring with one or two members of your group. Write the actual measurements in Column C.

4. Look at your chart. Answer this question: Which was greater, the group estimate (in Column B) or the actual measurement (in Column C)? Write the answer, either **B** or **C**, in Column D.

Measuring Things at School

	A Your Estimate	B Group Estimate	C Actual Measurement	D Is B Or C Greater?
How Many Inches?				
The Length of Your Thumb				
The Length of a New Pencil				
The Width of Your Math Book				
How Many Feet?				
The Height of Your Teacher's Chair				
The Height of the Class Chalkboard				
The Width of the Classroom Bookcase				
How Many Yards?				
The Distance from the Classroom Door to the Water Fountain				
The Distance from Your Classroom Door to the Fire Exit				
The Distance from Your Classroom Door to the Principal's Office				

Checkup

A. Listen to your teacher read two numbers. Write them on these lines. Ring the number that is greater in each pair.

1. _____ _____   2. _____ _____

3. _____ _____   4. _____ _____

B. Write the first nine even numbers in order.

____, ____, ____, ____, ____, ____, ____, ____, ____

C. Count by fives. Write the numbers.

____, ____, ____, ____, ____, ____, ____, ____, ____, ____

D. Here are heights of some mountains. Round each height to the nearest thousand feet.

1. Matterhorn   14,780 ft   About _____ ft

2. Rainier   14,408 ft   About _____ ft

3. Pikes Peak   14,108 ft   About _____ ft

E. Ring the best estimate.

1. A new pencil

   a. 6 feet   b. 6 inches   c. 6 yards

2. The length of a shoe

   a. 7 ft   b. 7 yd   c. 7 in.

3. The length of a classroom chalkboard

   a. 8 yards   b. 8 inches   c. 8 feet

4. The length of a schoolyard

   a. 225 ft   b. 225 in.   c. 225 yd

Addition

Look at the picture. Your teacher will tell you what to do.

Vocabulary for Addition

A. Read the following addition problems aloud to a friend. Your friend will write the problems in numbers on a sheet of paper. Then he or she will read the 10 problems to you, and you will write them in numbers on another sheet of paper.

Example: You read: "Eighteen plus thirteen equals thirty-one."

Your friend writes: 18 + 13 = 31   OR

$$\begin{array}{r} 18 \\ + 13 \\ \hline 31 \end{array}$$

1. Thirty-one plus thirteen equals forty-four.

2. Eighty-seven and twenty-two are a hundred nine.

3. Twenty-five added to fifty is seventy-five.

4. Four thousand five hundred added to two hundred fifty makes four thousand seven hundred fifty.

5. Three hundred twenty-four and eight hundred ninety-seven is one thousand two hundred twenty-one.

6. Ninety and ninety are one hundred eighty.

7. Nineteen and ninety is one hundred and nine.

8. Add four thousand to six thousand to get ten thousand.

9. The sum of twenty, thirty, and fifty is one hundred.

Check what you've written with your friend. Did you both write the same thing? Find and correct any mistakes.

B. To **add** is to put numbers or things together. You add to find out how many there are in all. 2 + 3 is an addition problem. You add to **solve** or **find the answer** to this problem.

The numbers that you add together are called **addends.** The answer is called the **sum.**

$$\begin{array}{r} 34 \quad \text{addend} \\ + 42 \quad \text{addend} \\ \hline 76 \quad \text{sum} \end{array}$$

To find the **total,** you add the numbers together. **Total** means the same as **sum. In all** and **altogether** also mean that you are thinking of numbers or things together.

Write a sentence or two telling what addition means. Use as many of these words as you can:

add	number	million	sum	math	greater
addend	equals	plus	solve	thousand	all
altogether	addition	total	one	two	answer

_____

_____

Problem Solving: The 5-Point Checklist

Use this 5-Point Checklist to help you solve problems.

QUESTION
DATA
PLAN
ANSWER
CHECK

1. Understand the QUESTION.
2. Find the needed DATA.
3. PLAN what to do.
4. Find the ANSWER.
5. CHECK the answer.

Understanding the Question

A. The first thing you do to solve a problem is understand the question. To understand the question, you can **find and rewrite the question.** Look for key words to help you.

   1. Find and underline the part of the word problem that asks the question. Usually the last sentence in the problem asks the question.
      In Washington School there are 10 classrooms on the first floor and 12 classrooms on the second floor. What is the total number of classrooms in Washington School?

   2. Find and ring the question words: **How much?, How many?,** or **What?**
      There are 5 fifth grade classrooms and 4 sixth grade classrooms. How many fifth and sixth grade classrooms are there in all?

      If you cannot find the question words, look for directions that tell you what to do: **Find the total number . . .** or **Find the sum.**
      There are 140 students in fifth grade, 120 students in sixth grade, and 100 students in seventh grade. Find the total number of students in these three grades.

   3. Rewrite the sentence that asks a question or gives a direction. Leave a blank where the answer will go.
      There are 25 students in Mr. Chávez's class and 32 students in Ms. Robertson's class. **How many students are there altogether in both classes?**

      There are ____ students altogether in both classes.

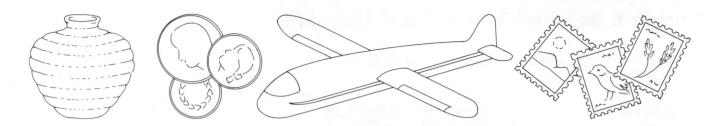

A **hobby** is something you do for fun. Some people collect stamps or coins as a hobby. Some people build model airplanes or cars as a hobby. Some people take photographs as a hobby.

Every year the Washington School has a Hobby Show. Students show examples of their hobbies, such as photographs, model planes, and comic book collections. All the students like to look at examples of their friends' hobbies.

B. Work with a friend on these problems. Find and ring the question words. Take turns reading each problem aloud. Underline the question. Rewrite the question as a statement. Leave a blank for the answer. Decide with your friend how to solve the problem. Write the answer in the blank.

1. Nadia's hobby is pottery. She uses clay to make things like bowls, cups, and plates. She has made 26 bowls, 32 cups, and 12 plates for the Hobby Show. How many pieces of pottery did Nadia make altogether for the Show?

   Rewrite the question: _____
   Solve the problem. Write the answer in the blank.

2. José's hobby is photography. He puts his best photos in an album. Find the total number of pages in José's album. He has already filled 37 pages with his photos. There are 23 empty pages left.

   Rewrite the question: _____
   Solve the problem. Write the answer in the blank.

3. Both María and Juan collect model airplanes. María has 35 models in her collection and Juan has 29 models in his. They decided to put together their two collections for the Hobby Show. How many model airplanes do María and Juan have in the Hobby Show?

   Rewrite the question: _____
   Solve the problem. Write the answer in the blank.

4. Nobuo collects old comic books. He took 13 *Superman,* 18 *Captain Marvel,* and 15 *Spiderman* comic books to the Hobby Show. He has 284 more comic books at home. How many comic books does he have in all?

   Rewrite the question: _____
   Solve the problem. Write the answer in the blank.

Finding the Needed Data

A. Most word problems contain numbers or data. There are two steps in finding the data you need to solve a problem:

1. Read the question carefully. Ask yourself: "What numbers do I need to solve the problem?"
2. Ring the key numbers—the numbers you need to solve the problem.

QUESTION
DATA
PLAN
ANSWER
CHECK

Now you are going to practice finding the needed data. First read the sample problem carefully. The key numbers are ringed. Write the answer for the sample problem.

Sample problem: Juan had ③ records. Answer: _____
Yesterday was Juan's birthday. He was 14. His uncle gave him ② records for his birthday. Now how many records does Juan have altogether?

B. Do the problems below. Work by yourself.

1. Read each problem carefully.
2. Circle the key numbers.
3. Solve the problem.
4. Write the answers in the blanks.

1. Tina bought 2 records and some record cleaner for $15. María bought 1 record for $6. How much money did Tina and María spend altogether?   Answer: _____

2. Nu Trinh had 9 records. She bought 6 records on sale for $20. What is the total number of records in Nu Trinh's collection?   Answer: _____

3. María has 17 records in her collection. Juan has 12 in his, and José has 5 in his. How many records do María and José have together?   Answer: _____

4. The clerk in the record store sold 129 records before 12:00. She sold 87 records in the afternoon. At 6:00 she went home. How many records did she sell that day?   Answer: _____

C. Now work with two or three friends to check your work. (You may use a calculator.) Do you agree on the data you needed?

How many correct answers did you have? _____ Which

problems were easy? _____ Difficult? _____ Look again at the problems that were difficult. Decide why they were difficult. Is the math too hard? Are the words too hard? Write a sentence that tells what was difficult.

Plan What to Do: Draw a Picture

A. Sometimes drawing a picture makes it easier to know what to do. This is part of planning. You can make the picture any way you want.

   Look at the sample problem. The question is underlined and rewritten, with space for the answer. Finish the problem and write the answer. Then do the four problems below.

> **Sample problem:** A large circus tent has 5 poles from one end to the other end. The poles are 20 feet apart. **How long is the tent?**

Rewrite the question: <u>The tent is _____ long.</u>

20 ft   20 ft   20 ft   20 ft

QUESTION
DATA
PLAN
ANSWER
CHECK

(Draw your picture here.)

B. Do these problems. Work by yourself.

1. Juan has a model train set with one engine, 3 passenger cars, and a caboose. How many cars are on the train altogether?

   Rewrite the question: _____

2. Nu Trinh is 56 inches tall. Her mother is 4 inches taller. How tall is Nu Trinh's mother?

   Rewrite the question: _____

3. María had 3 posters. On her birthday, each of her four friends gave her a poster. How many posters did she have altogether?

   Rewrite the question: _____

4. Tran was reading a book called *The Planet Earth.* He learned that there are 2 planets that are closer to the sun than the Earth is. Six planets are farther from the sun than the Earth is. How many planets are in the solar system altogether?

   Rewrite the question: _____

C. Share your drawings with a small group of friends. Did they have the same kinds of drawings as you? Did your pictures help you understand the problems? How?

Write Your Own Problems

Now it is your turn to write your own addition word problems. Follow these steps:

1. Organize your ideas.
2. Write a word problem.
3. Read your problem to two friends. Solve each other's problems.
4. Check your answers with your friends.

A. Organize your ideas. First choose an addition equation.

   Examples: 33 + 82 = 115   264 + 367 = 631

   Then think of a story to go with the equation.

   Examples: My brother has 33 records and my dad has 82.
   José had 264 stamps in his stamp collection. His
   uncle gave him 367 more stamps.

B. Write a word problem. First write the story you thought about.
   Then write a question to go with the story. Remember to use
   words that tell what math operation to use.

   Examples: How many records do they have altogether?
   How many stamps does José have in all?

C. Now try writing other addition word problems.

Problem 1

Addition equation: _____

Story: _____

_____

Question: _____

Problem 2

Addition equation: _____

Story: _____

_____

Question: _____

Problem 3

Addition equation: _____

Story: _____

_____

Question: _____

Problem 4

Addition equation: _____

Story: _____

_____

Question: _____

D. Sit with two friends. Take turns reading your problems and solving them. Read your problems aloud. As you read, your friends will write down the important numbers. Then they will solve your problems. When it is your turn to solve their problems, remember to use the 5-Point Checklist.

Use this space to solve your friends' problems.

E. Now work with your two friends to check your work. (You may use a calculator.) How many correct answers did you

have? _____ Which problems were easy? _____

Difficult? _____ Look again at the problems that were difficult. Decide why they were difficult. Is the math too hard? Are the words too hard? Write a sentence that tells what was difficult.

_____

Finding the Perimeter

A. The distance around the outside of an object is called the **perimeter.** The object can be a picture, a bulletin board, a park, a baseball field, or anything that has sides. It can also be a figure such as a square, a triangle, or a rectangle. We use measurements of length and distance to find the perimeter.

Rule: To find the perimeter of an object or figure, add the lengths of its sides together.

Example: José ran around the outside of a field. How far did he run?

```
            120 yd
    ┌──────────────────────┐
    │                      │
    │                      │ 45 yd
    │                      │
    └──────────────────────┘
```

We add the lengths of the four sides to get the total distance around the field:

```
   120
    45
   120
 + 45
  ────
   330        José ran 330 yards.
```

B. Now your teacher will read some perimeter problems. Draw a picture of the object or figure. Write down the numbers that tell the lengths of the sides. Then solve the problem.

1. Draw a picture and write the numbers:

Write the answer: _____

2. Draw a picture and write the numbers:

Write the answer: _____

3. Draw a picture and write the numbers:

Write the answer: _____

Making Reasonable Choices

We can use what we know about perimeter, measurement, and addition to make reasonable choices. A **reasonable choice** is a choice that makes sense. It fits the problem.

A. Work with a friend. Follow these steps as you solve the problems below, on measurement and addition.

1. Read each problem aloud.
2. Draw a picture.
3. Ring your choice for the answer.

1. Mrs. Wong had a rectangular garden plot that was 3 ft wide and 10 ft long. She wanted to put an 8-inch-high border fence along its edges. About how much fencing material did she need?

   a. 21 ft   **b.** 26 yds   **c.** 27 ft

2. Mr. García had a rectangular piece of plywood that was 6 ft long and 3 ft wide. He wanted to put model railroad track around its edge. Each piece of track was 1 ft long plus 4 corner pieces. About how many foot-long pieces should he buy?

   a. 9   **b.** 18   **c.** 21

3. Helen made a baby blanket that needed a border. The rectangular blanket was 34 inches wide and 44 inches long. She had 140 inches of border material. How much did she have?

   a. not enough   **b.** the exact amount   **c.** more than she needed

4. Warren had a picture to frame. The picture was 5 inches wide and 7 inches long. He needed tape to put around the edges of the picture. He had 30 inches of tape. How much did he have?

   a. not enough   **b.** the exact amount   **c.** more than he needed

B. Compare your choices with two other friends. Give the reasons for your choices. Do you agree? _____

C. Write a problem for your friends to solve. Try to give them some choices.

_____

_____

Checkup

A. Write the addition equations as your teacher reads them aloud.

1. _____ **2.** _____ **3.** _____

4. _____ **5.** _____ **6.** _____

B. Rewrite the question in each of the following problems. Then write the key words and the operation. Write the answer on line **a.**

1. María and Ben traveled 13 miles down Shoelace Creek in a boat. They stopped for lunch and then went 5 miles more. How many miles did they go in the boat altogether?

a. Rewrite the question: _____

b. Write the key word: _____ **c.** Write the operation: _____

2. Tran traveled 12 miles to the museum. Then he traveled 4 miles to the park. How many miles did he travel in all?

a. Rewrite the question: _____

b. Write the key word: _____ **c.** Write the operation: _____

C. Circle the key numbers you need to solve each of the following problems. Then find the answer.

1. James and Su-Linn began collecting stamps when he was 8 and she was 10 years old. He has 23 stamps and she has 35. They have been collecting for 6 months. How many stamps do they have?

(Be sure to circle the key numbers.) Answer: _____

2. Eduardo weighs 74 pounds. His mother weighs 122 pounds. His little sister weighs 23 pounds. How many pounds do Eduardo and his sister weigh together?

(Be sure to circle the key numbers.) Answer: _____

D. Draw a picture for the problem below. Then find the answer. There was one slot in the bicycle rack for Donna to put her bike. There were 6 bikes to the right of Donna's. Hers was in the middle of the rack. How many bikes were in the bicycle rack in all? Answer: _____

Subtraction

Look at the picture. Your teacher will tell you what to do.

Vocabulary for Subtraction

Subtraction is a math operation. You subtract to find out **what the difference is** between two numbers. Other ways to ask what the difference is are **how many more?, how much more?, how much is left?,** and **how much remains?**

A. Work with a friend. Read these examples aloud:

1. Lisa has 7 plants.   Rosa has 5 plants.

THINK   **How many more** plants does Lisa have than Rosa?

I subtract 5 from 7.

$$\begin{array}{r} 7 \\ -\ 5 \\ \hline 2 \end{array}$$

Lisa has 2 more plants than Rosa.

2. Tran has [ONE] [ONE] [ONE]

He spends [ONE] [ONE]

THINK   **How much money** is left?

I subtract $2.00 from $3.00.

$$\begin{array}{r} \$3.00 \\ -\ 2.00 \\ \hline \$1.00 \end{array}$$

Tran has $1.00 left.

3. María has [TEN]

Sam has [FIVE]

THINK   **How much more** money does María have than Sam?

I subtract $5.00 from $10.00.

$$\begin{array}{r} \$10.00 \\ -\ 5.00 \\ \hline \$\ \ 5.00 \end{array}$$

María has $5.00 more than Sam.

4. ⑥   ④

THINK   What is the difference between 6 and 4?

I subtract 4 from 6.

$$\begin{array}{r} 6 \\ -\ 4 \\ \hline 2 \end{array}$$

The difference between 6 and 4 is 2.

B. Work with a friend. Write short sentences that use the vocabulary for subtraction. Use some numbers in each sentence. The first one is done for you.

less than   more than   difference
fewer than   greater than   left
lighter than   heavier than   remaining
shorter than   taller than   farther than

1. _25 is 5 less than 30._

2. _____

3. _____

4. _____

5. _____

6. _____

7. _____

8. _____

9. _____

10. _____

11. _____

12. _____

C. Now look over your sentences and correct them if you need to. Sit with two or three friends. Read your sentences aloud. Then listen to your friends read their sentences to you. Write the equations that you hear.

1. _____   2. _____   3. _____

4. _____   5. _____   6. _____

7. _____   8. _____   9. _____

10. _____   11. _____   12. _____

Addition and Subtraction Properties

The math operations **addition** and **subtraction** are related. A **Fact Family** shows how.

FACT FAMILY	FACT FAMILY	FACT FAMILY
2 + 5 = 7	7 + 9 = 16	8 + 0 = 8
5 + 2 = 7	9 + 7 = 16	0 + 8 = 8
7 − 5 = 2	16 − 9 = 7	8 − 0 = 8
7 − 2 = 5	16 − 7 = 9	8 − 8 = 0

Fact Families help you understand the **properties,** or rules, for addition and subtraction.

A. Use the Fact Families above to answer these questions. Work by yourself.

1. In addition, do you get the same answer when you change the order of the addends? _____

2. In subtraction, can you change the order of the two numbers and still get the same answer? _____

3. What happens when you add 0 to a number? _____

4. What happens when you subtract 0 from a number? _____

5. What happens when you subtract a number from the same number? _____

B. Work with a friend. Take turns reading the questions or statements aloud. Decide whether each one is talking about **addition, subtraction,** or **both** operations. Try some examples with numbers to help you decide. Then write + for addition, − for subtraction, or **both** on the line following the question or statement.

1. You can change the order of the numbers, but the answer will be the same. _____

2. If two numbers (other than zero) are the same and the answer is 0, what operation did you use? _____

3. When you have to write the larger of two numbers first, which operation are you going to use? _____

4. When you have to find the total number, which operation do you use? _____

5. If a problem asks how much money is left after you spend some, which operation should you use? _____

C. Now check your work with a different friend. Do you have the same answer? If not, read the question again and talk about it. Decide together on the correct answer.

Problem Solving: Subtraction

One of the things you do to solve a problem is to plan what to do. For example, you can:

1. Draw a picture.
2. Choose the math operation.

QUESTION
DATA
PLAN
ANSWER
CHECK

Draw a Picture

A. Remember that drawing a picture sometimes makes it easier to know what to do. You can make the picture any way you want. When you read a problem, follow these steps:

1. Find and underline the question.
2. Rewrite the question. Leave a blank for the answer.
3. Draw a picture. Choose the correct math operation.
4. Solve the problem. Write the answer.

1. Luis and Lin went shopping for school supplies. Luis bought 5 pencils and 2 pens. Lin bought 15 pencils and 1 tablet. Who bought more pencils? How many more?

(Draw your picture here.)

Rewrite the question: _____
Solve the problem. Fill in the answer.

2. Twelve children were playing soccer. Seven children went home. How many remained?

Rewrite the question: _____
Solve the problem. Fill in the answer.

3. Sara has to peel 8 potatoes. She has already peeled 3. How many more potatoes does she have to peel?

Rewrite the question: _____
Solve the problem. Fill in the answer.

B. Share your drawings with a small group of friends. Talk about these questions with your friends. Then write your answers.

1. Did your friends have the same kind of drawings as you? _____

2. What pictures were different? _____

3. Did drawing pictures help you understand the problems? _____

4. How did drawing pictures help you? _____

Choose the Math Operation

A. In problem solving, you often need to choose the correct math operation. Two math operations are addition and subtraction. You can tell which operation to use by looking for the **key words.** Some key words that tell you to subtract were given on page 27. Study them. Then follow these steps:

1. Find and circle the key words in each problem.
2. Rewrite the question. Leave a blank for the answer.
3. Choose the operation and solve the problem.
4. Write your answer in your rewritten question.
5. Check the answers with a friend when you are done.

The Treasure Hunt

You have been given a map to a buried treasure. The treasure is 1,097 yards from where you are now standing at Tall Tree if you cross Bad Bridge. But Bad Bridge is dangerous and might break. You can also get to the treasure by following the trail to Round Rock. The distance to the treasure by going to Round Rock is 2,605 yards.

Round Rock

Trail

Tall Tree

Bad Bridge

1. How much farther is it to go past Round Rock than it is to go across Bad Bridge? (Hint: Write the distances on the map.)

   Rewrite the question: _____
   Solve the problem. Fill in the answer.

2. It is 623 yards from Round Rock to the treasure. It is 458 yards from Bad Bridge to the treasure. How much closer is the treasure to Round Rock than to Bad Bridge? (Hint: Find the distances on the map.)

   Rewrite the question: _____
   Solve the problem. Fill in the answer.

3. Roan decides to walk to Bad Bridge to make sure that it is really too dangerous to cross. How far does he have to walk to get to Bad Bridge from Tall Tree?

   Rewrite the question: _____
   Solve the problem. Fill in the answer.

B. Now check your answers with your friends. Which problems

were difficult? _____

Why were they difficult? _____

Problem Solving: Addition or Subtraction?

You need to choose the correct **math operation** to solve a word problem. In the problems on this page you need to choose either **addition** or **subtraction** to solve the problem.
Directions:

1. Find and ring the key words in each problem.
2. Rewrite the question. Leave a blank for the answer.
3. Choose the math operation: addition or subtraction?
4. Solve the problem.
5. Write your answer in the blank space in your rewritten question.
6. Check your answers.

A. These things are for sale in a catalog. Use the data for these problems.

$2.49

$7.25 $7.95

$8.20

$6.50

1. How much are the calculator and flashlight together?

   a. Ring the key words.

   b. Rewrite the question. _____

   c. What is the operation? _____
   d. Solve the problem. Write your answer in the space.

2. How much more is the clock than the flashlight?

   a. Ring the key words.

   b. Rewrite the question. _____

   c. What is the operation? _____
   d. Solve the problem. Write your answer in the space.

3. How much less is the calculator than the clock?

   a. Ring the key words.

   b. Rewrite the question. _____

   c. What is the operation? _____
   d. Solve the problem. Write your answer in the space.

4. What is the total price for the radio, the flashlight, and the game?

   a. Ring the key words.

   b. Rewrite the question. _____

   c. What is the operation? _____
   d. Solve the problem. Write your answer in the space.

B. Now check your answers with those of one or two friends.

   Which problems were difficult? _____ Why? _____

Write Your Own Problems

Now it is your turn to write subtraction and addition problems. Follow these steps:

1. Read the data.
2. Organize your ideas. Choose an addition or subtraction fact, using the data. Then think of a story.
3. Draw a picture of your problem.
4. Write a word problem.
5. Read your problem to two friends. Solve each other's problems.
6. Check your answers with your friends.

A. Read the data.

A skyscraper is a very tall building. Here are the names and heights of some famous skyscrapers:

NAME	CITY	HEIGHT
John Hancock Center	Chicago	337 meters
Standard Oil Building	Chicago	346 meters
Empire State Building	New York City	381 meters
World Trade Center	New York City	411 meters
Sears Tower	Chicago	443 meters

B. Now write your problems.

Problem 1

Addition or subtraction fact: _____
Picture:

Story: _____

_____

Question: _____

Problem 2

Addition or subtraction fact: _____
Picture:

Story: _____

_____

Question: _____

Problem 3

Addition or subtraction fact: _____
Picture:
        Story: _____

_____

Question: _____

Problem 4

Addition or subtraction fact: _____
Picture:
        Story: _____

_____

Question: _____

C. Sit with two friends. Take turns reading your problems and solving them. Read your problems aloud. As you read, your friends will write down the important numbers. Then they will solve your problems. Remember to use the 5-Point Checklist.
    Use this space to solve your friends' problems.

D. Now work with your two friends to check your work. (You may use a calculator.) How many correct answers did

you have? _____ Which problems were

easy? _____ Difficult? _____ Look again at the problems that were difficult. Decide why they were difficult. Is the math too hard? Are the words too hard? Write a sentence that tells what was difficult.

_____

Problem Solving: Estimation

One way to check your answer to a problem is to **estimate,** or to guess about how many the answer is. You guess about how many the answer is by using rounded numbers. For example:

nearest ten

30 _ _ 33 _ _ _ _ _ _ 40

33 is closer to 30 than to 40. Round 33 to the nearest 10.

33 rounded is 30.

nearest ten

120 _ _ _ _ 125 _ _ _ _ 130

125 is halfway between 120 and 130. Round up when the number is halfway or more.

125 rounded is 130.

nearest hundred

1,600 _ _ 1,638 _ _ _ _ _ _ 1,700

1,638 is closer to 1,600 than to 1,700. Round 1,638 to the nearest hundred.

1,638 rounded is 1,600.

nearest hundred

2,100 _ _ _ _ 2,152 _ _ _ _ 2,200

2,152 is closer to 2,200 than to 2,100. Round 2,152 to the nearest hundred.

2,152 rounded is 2,200.

A. Use estimation to decide which answer is closest to being correct in the following problems. You must be able to choose the correct operation (addition or subtraction) to get the answer. Ring the correct answer.

1. Julio and his father exercise together. Julio can do 35 situps and his father can do 83. About how many more situps can Julio's father do than Julio?

a. 10   **b.** 20   **c.** 30   **d.** 40

2. Mai and her older brother tried to see how far they could run in 5 minutes. She ran 882 yards and he ran 1,173 yards. About how many yards farther did her brother run than Mai?

a. 200   **b.** 300   **c.** 400   **d.** 500

3. Juanita and her sister swam for 27 minutes. Then they jogged for 19 minutes. About how many minutes did they exercise?

a. 40   **b.** 50   **c.** 60   **d.** 70

4. In the baseball-throwing contest, the school team of three students threw for 45 yards, 52 yards, and 57 yards. About how many total yards did they throw the ball?

a. 160   **b.** 170   **c.** 180   **d.** 190

B. Now check all your answers with a friend. Did your friend get the same answers as you? _____ If not, take turns describing why you gave the answer that you did.

Graphs

A **graph** is a picture that gives you information about numbers. In a graph you must know what information is being presented.

A. Work in a group with three friends. Complete the sample problem below. Then do the other problems.

Sample Problem. Count the number of boys and girls in your class. Enter the information here:

_____ Girls   _____ Boys

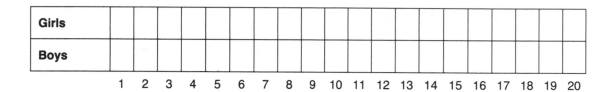

B. Now place a mark on the graph at the number of girls you counted. Fill in the bar for "girls" up to the mark you just made. Use pencil or pen. Do the same for "boys." This bar graph shows:

☐ how many boys and girls there are in your class.
☐ if there are more girls than boys, or more boys than girls.

C. Write a sentence that describes what you see on the graph you made. For example, if there were 14 girls and 12 boys, you would write:

There are 14 girls and 12 boys in my class. There are more girls than there are boys.

_____

_____

_____

D. Compare your graph with a friend's. Do the graphs look alike? Why?

_____

_____

Kim's grade in school voted to see which animal would be the name for their soccer team. Kim made a tally chart to show how many votes each type of animal received.

ANIMALS   **TALLIES OF ANIMALS**

Bulldogs   ⵏⵜ ⵏⵜ ⵏⵜ ⵏⵜ

Chicken Hawks   ⵏⵜ ⵏⵜ ‖

Panthers   ⵏⵜ ⵏⵜ ⵏⵜ ‖

E. Fill in the bars on the graph. Show how many votes each animal received for team name. Notice that on this graph you must count in fives and fill in **about** where the top of the bar goes. When you have completed the graph, answer the following questions. (Use complete sentences.)

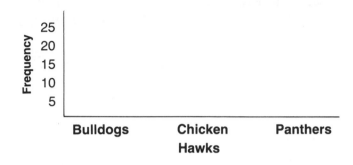

1. How many votes were for Bulldogs?

_____

2. How many votes were for Chicken Hawks?

_____

3. How many votes were for Panthers?

_____

4. Which animal had the most votes?

_____

5. Which animal received the fewest votes?

_____

F. Read your answers to a friend. Listen while your friend reads to you. Do your answers agree? _____

Mr. and Mrs. Pérez watched a big race over the weekend. They wanted to find out how long it took most of the runners to finish. They wrote down all of the finishing times in minutes. Next they made four groups of times. They used the groups to make a tally chart. Here are the finishing times. They should be read in minutes and seconds. The first one is "10 minutes, 7 seconds."

Runner

1. 10:07	2. 9:48	3. 10:13	4. 9:46
5. 10:08	6. 9:55	7. 9:45	8. 10:04
9. 10:15	10. 10:20	11. 9:58	12. 10:16
13. 10:10	14. 10:12	15. 10:18	

G. Now you make the tally chart for Mai.

TIMES	TALLIES OF TIMES
9:45–10:00	_____
10:01–10:15	_____
10:16–10:30	_____

H. Use your tallies to complete the bar graph.

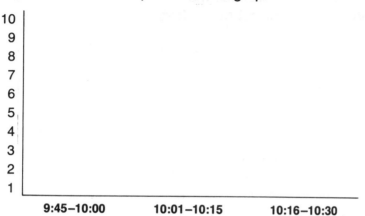

I. Answer these questions about the bar graph. (Use complete sentences.)

1. How many runners finished in 10:00 minutes?

_____

2. How many runners finished in 10 minutes and 15 seconds?

_____

3. How many runners finished in 10 minutes and 30 seconds?

_____

4. Did most of the runners finish in 10 minutes?

_____

J. Now check all of your answers with a friend. Look at the graph your friend made. Do the graphs look the same? _____ Are any of them different? _____ How are they different? _____ Why do you think that the graphs are different? _____

_____

Now look at your friend's answers to each question. Are your answers the same? _____ Are any of them different? _____ How are they different? _____

Checkup

A. Write the subtraction equations for each problem your teacher reads aloud.

    1. _____   **2.** _____

    3. _____   **4.** _____

B. Decide what operation each of the following sentences make you think of. Write **+ for addition, − for subtraction,** or **both.**

    1. You can change the order of the numbers and the answer will

        be the same. _____

    2. The answer in this operation is called "the difference." _____

C. In each of the following problems, ring the key numbers and words. Then rewrite the question and solve the problem. Write your answer in the blank in your rewritten question.

    1. Kevin is 58 inches tall. His brother is 66 inches tall. His dad is 73 inches tall. How much taller than Kevin is Kevin's dad?

        Rewrite the question _____

    2. A wristwatch costs $12.98. A calculator costs $7.87. What is the difference between the two?

        Rewrite the question _____

D. Ring the best answer.

    1. Juan ran for 27 minutes. Then he walked for 41 minutes. About how many more minutes did he walk than run?

        a. 10 minutes   **b.** 20 minutes   **c.** 30 minutes

E. Use the graph. Answer each question in a full sentence.

    1. About how many students ride the bus to school?

        _____

        _____

    2. About how many fewer students ride the bus to school than walk?

        _____

        _____

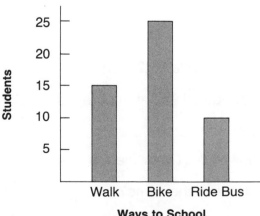

Ways to School

Multiplication

Look at the picture. Your teacher will tell you what to do.

Vocabulary for Multiplication

A. Work with two or three friends. Talk about each math word and then match the math word with its definition. Write the correct letter next to each word.

1. factor _____

2. product _____

3. times _____

4. multiplication _____

5. equation _____

a. a number sentence that uses the equal sign ($=$).

b. to put together equal sets of numbers (or to add a number to itself one or more times)

c. the result of multiplication

d. multiply

e. a number that you multiply by another number

B. Work with a friend. Your friend should cover the equations left of the dotted line. As you read each equation aloud, your friend writes it in his or her book. For example, you read: "Five times ten is fifty." Your friend writes: $5 \times 10 = 50$.

1. $3 \times 2 = 6$ _____

2. $4 \times 5 = 20$ _____

3. $6 \times 3 = 18$ _____

4. $5 \times 10 = 50$ _____

5. $6 \times 4 = 24$ _____

6. $7 \times 3 = 21$ _____

7. $8 \times 2 = 16$ _____

8. $9 \times 6 = 54$ _____

9. $10 \times 0 = 0$ _____

10. $10 \times 10 = 100$ _____

Now you cover the equations. Your friend will read the equations. Listen and write them in your book.

C. Uncover the equations. Check your work. How many right answers did you have? _____ How many wrong answers did you have? _____ What numbers are difficult to understand? _____

Multiplication Properties

Multiplication has three important properties, or rules. Read each rule. Then read the examples aloud.

1. The **Order Property.** You can change the order of the factors. The product will be the same.

    $3 \times 2 = 6$ $12 \times 10 = 120$

    $2 \times 3 = 6$ $10 \times 12 = 120$

2. The **One Property.** If **1** is a factor, the product will be the same as the other factor.

    $5 \times 1 = 5$ $999 \times 1 = 999$

    $15 \times 1 = 15$ $48 \times 1 = 48$

3. The **Zero Property.** If **0** is a factor, the product will be 0.

    $4 \times 0 = 0$ $562 \times 0 = 0$

    $11 \times 0 = 0$ $76 \times 0 = 0$

Do you remember the addition properties? Review them:

1. The **Order Property.** You can change the order of the addends. The sum will be the same.

    $5 + 3 = 8$ $72 + 10 + 5 = 87$

    $3 + 5 = 8$ $10 + 5 + 72 = 87$

2. The **Zero Property.** If one addend is **0,** the sum will be the same as the other addend.

    $11 + 0 = 11$ $587 + 0 = 587$

    $0 + 29 = 29$ $2134 + 0 = 2134$

A. Work with a friend. Take turns reading the questions or statements aloud. Decide whether each one is talking about **multiplication, addition,** or **both** operations. Then write × **for multiplication, + for addition,** or **both.**

1. _____ You can change the order of the numbers, and the answer will be the same.

2. _____ If one number is 0 and the answer is 0, what operation did you use?

3. _____ If one of the two numbers is 0 and the answer is the same as the other number, did you use addition or multiplication?

4. _____ If one number is 1 and the answer is the same as one of the other numbers, what operation did you use?

5. _____ If one number is 1 and the answer is one more than the other number, what operation did you use?

Understanding Multiplication

Aleta planted some rows of vegetables. She had 3 rows of corn with 4 plants in each row. She told her sister that she had 12 corn plants in all.

George planted some vegetables too. He had 4 rows of corn with 3 plants in each row. He also said that he had 12 corn plants in all.

　　Both Aleta and George wrote some multiplication expressions to show how many plants they had.

Aleta wrote:

$$\begin{array}{r} 4 \\ \times\ 3 \\ \hline \end{array}$$ 3 fours = 12
　　　　3 × 4 = 12

George wrote:

$$\begin{array}{r} 3 \\ \times\ 4 \\ \hline \end{array}$$ 4 threes = 12
　　　　4 × 3 = 12

"Three times four equals twelve."

"Four times three equals twelve."

A. Read what Aleta and George wrote. How are these expressions alike? Aleta and George planted other vegetables also. Work with a friend on these problems. Read and talk about each problem together. Use multiplication and fill in the blanks. Say each expression aloud.

1.

2 fours = _____

2 × 4 = _____

two times _____ equals _____

2.

4 twos = _____

4 × 2 = _____

four times _____ equals _____

3.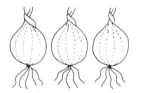

2 threes = _____

2 × 3 = _____

$$\begin{array}{r} 2 \\ \times\ 3 \\ \hline \end{array}$$

4.

3 twos = _____

3 × 2 = _____

$$\begin{array}{r} 2 \\ \times\ 3 \\ \hline \end{array}$$

5.

2 × 5 =

_____ times _____ equals _____

6.

5 × 2 =

five _____ two _____ _____

7.

$2 \times 3 =$ _____

two _____ _____ equals _____

8.

$3 \times 2 =$

three _____ _____ _____

_____

9.

$$\begin{array}{r} 3 \\ \times\ 3 \\ \hline 9 \end{array}$$

_____

_____

10.

$$\begin{array}{r} 2 \\ \times\ 2 \\ \hline 4 \end{array}$$

_____

_____

B. Sit with two or three other friends to check your work. How often

did you know how to solve the problem? _____ How many

correct answers did you have? _____ Look again at the
problems that were difficult. Decide why they were difficult. Did

you know how to solve them? _____

Did you get the wrong answer anyway? _____

C. Give information below. Then review the correct answers with a
friend.

1. Words in the directions and problems that I did not know:

_____

_____

2. Problems I know how to solve but my answer was wrong:

_____

3. Problems I did not know how to solve:

_____

Problem Solving: Multiplication

Work with a friend on these problems. Take turns reading the problems aloud. Follow these two steps:

1. Tell your friend how to find the data that you need to solve the problem. Then ring these key numbers.
2. Now tell your friend how to find the words that tell which operation to use. Then ring these key words.

QUESTION
DATA
PLAN
ANSWER
CHECK

A. Ring the key numbers and key words. Tell your friend how you know what to do. Solve each problem and tell whether you used addition or multiplication.

1. Frank has a coin folder for nickels. It has spaces for 16 nickels on each page. Frank has filled 3 pages. How many

   nickels does he have altogether? _____

2. Sumi has collected 78 dimes. Her coin folder holds 20 dimes on a page and has 3 pages. What is the total number of

   dimes she can put in the folder? _____

3. Rob has collected 17 quarters, 32 dimes, and 85 pennies.

   How many coins does he have altogether? _____

4. Tran has a coin folder for quarters. He has filled each of 4 pages with 16 quarters. How many quarters does he have

   in all? _____

5. Margaret has a coin folder for dimes. She put 18 dimes on one page, 15 dimes on the next, and 119 dimes on the third

   page. How many coins does she have in all? _____

6. María has collected 125 pennies, 34 nickels, 48 dimes, and 25 quarters. What is the total number of coins that she

   has? _____

B. Write here the key words that you circled for multiplication and for addition:

Multiplication   Addition

_____   _____

_____   _____

_____   _____

What can you say about these words? _____

Compare your key words with a friend's. Do they agree? _____

Multiplying with Money

Lin Quong owns the Party Planner store in San Francisco. She is ordering some party favors to sell in her store. She has filled out the order form to show how many of each item she wants.

QUESTION
DATA
PLAN
ANSWER
CHECK

PARTY PLANNER
Order Form

Item	Catalog Number	Quantity	Price of Each	Total Price
1. Balloons (pkg)	3-41	7	$1.29	_____
2. Top Hat	3-22	3	5.23	_____
3. Party Hats	3-24	8	2.31	_____
4. Decorations	7-53	2	1.42	_____
5. Candles (pkg)	7-48	5	.75	_____
6. Horns	6-33	9	.89	_____
7. Noisemakers	6-82	3	1.78	_____
8. Masks	4-12	6	4.19	_____
			TOTAL	_____

A. Use the order form to answer these questions:

   1. How much do 6 masks cost? _____

   2. How much do 5 packages of candles cost? _____

   3. What is the cost of 1 noisemaker and 1 horn? _____

   4. How much more does a top hat cost than a party hat? _____

   5. What is the total cost of the order? _____

B. Check your answers with two friends.

Write Your Own Problems

Now it is your turn to write multiplication problems. Follow these steps:

1. Read the data.
2. Organize your ideas. Think of a story.
3. Draw a picture of your problem, or make a table.
4. Write a word problem.
5. Read your problem to two friends. Solve each other's problems.
6. Check your answers with your friends.

A. Read the data.

In your word problems, pretend you are planning for a picnic. For each problem, use different numbers of people, or different numbers of items from the list. Fill in the number, price, and total for the items you want; then add to get the total. Add your own items if you wish.

Item	Number	Price	Total
Hot dog buns	_ _ _ _	_ _ _ _	_ _ _ _
Hot dogs	_ _ _ _	_ _ _ _	_ _ _ _
Cheese (10 slices)	_ _ _ _	_ _ _ _	_ _ _ _
Paper cups	_ _ _ _	_ _ _ _	_ _ _ _
Lemonade mix	_ _ _ _	_ _ _ _	_ _ _ _
_ _ _ _ _ _	_ _ _ _	_ _ _ _	_ _ _ _
_ _ _ _ _ _	_ _ _ _	_ _ _ _	_ _ _ _

B. Now write your problems.

Problem 1

Multiplication fact: _____

Picture:

        Story: _____

_____

Question: _____

Problem 2

Multiplication fact: _____

Picture:

        Story: _____

_____

Question: _____

Problem 3

Multiplication fact: _____

Picture:

       Story: _____

_____

Question: _____

Problem 4

Multiplication fact: _____

Picture:

       Story: _____

_____

Question: _____

C. Sit with two friends. Take turns reading your problems and solving them. Read your problems aloud. As you read, your friends will write down the important numbers. Then they will solve your problems.

    Use this space to solve your friends' problems.

D. Work with your two friends to check your work. (You may use a calculator.) How many correct answers did you have? _____

Which problems were easy? _____ Difficult? _____ Look again at the problems that were difficult. Decide why they were difficult. Is the math too hard? Are the words too hard? Write a sentence that tells what was difficult.

_____

Estimating Products

When you want an answer that is **close** to the exact answer, you estimate. To estimate, you round the numbers. Then you do the math operation in your head.

Example: A canary's heart beats about 22 times faster than an elephant's heart. How many times does a canary's heart beat in 1 minute?

Animal Heartbeats in 1 Minute	
Animal	**Heartbeat**
Camel	31
Elephant	33
Giraffe	66
Lion	44

A. Use the table to find the data. Think of the problem: 22 × 33. Round the numbers: 20 × 30.
Multiply in your head: 20 × 30 = 600.

Write the answer: _____

B. Work with one or two friends. Use the data about animal heartbeats in the table to estimate other animal heartbeats. Write a complete sentence for each answer.

   1. A bat's heart beats about 23 times faster than an elephant's heart. About how many times does a bat's heart beat in 1 minute?

   a. Round the numbers: _____   **b.** Multiply in your head: _____

   c. Write the answer: _____

   2. A mouse's heart beats about 12 times faster than a lion's heart. About how fast does a mouse's heart beat in 1 minute?

   a. Round the numbers: _____   **b.** Multiply in your head: _____

   c. Write the answer: _____

   3. A canary's heart beats about 11 times faster than a giraffe's heart. About how fast does a canary's heart beat in 1 minute?

   a. Round the numbers: _____   **b.** Multiply in your head: _____

   c. Write the answer: _____

   4. A rat's heart beats about 11 times faster than a camel's heart. About how fast does a rat's heart beat in 1 minute?

   a. Round the numbers: _____   **b.** Multiply in your head: _____

   c. Write the answer: _____

C. What key word in each problem tells you that only an estimate is

wanted? _____

Measurement: Area

We use square units to find the **area** of a shape.

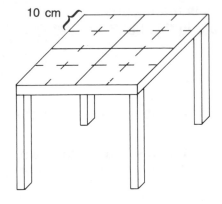

A. Work with three friends. Use paper square units to measure shapes in your classroom. Look at the example:
The area of the table is 4 construction paper units. Each construction paper square that is 10 centimeters on a side is 100 square centimeters. So 4 of these squares equal 400 square centimeters.

one 10 × 10 square = 100 square centimeters
four 10 × 10 squares = 400 square centimeters

The area of the table is 1600 square centimeters.

B. Work with your group to measure the following areas in your classroom. (If the areas that you measure are not a complete number of paper squares, round the measurement to the nearest whole unit.)

1. A student desk or table.

_____ construction paper units

_____ square centimeters

2. Your teacher's desk.

_____ construction paper units

_____ square centimeters

3. A bulletin board.

_____ construction paper units

_____ square centimeters

4. A book.

_____ construction paper units

_____ square centimeters

5. You choose! Name of object _____

_____ construction paper units

_____ square centimeters

C. Check your group's work with that of another group.

Measurement: Volume

Volume is the number of cubic units that you need to fill a space.

How many cubes
are in this box?

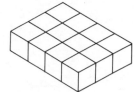

3 rows of cubes

4 in each row

1 layer of cubes

To find the volume of the box: 3 × 4 × 1 = 12 cubes in the box.

A. Work with a friend. Take turns reading and following the
directions for each problem. Do these steps:

1. One person reads the directions for drawing a picture.
2. The other person draws the picture.
3. Discuss and write the equation.
4. Write the volume of each box in cubic units.

1. Directions: Draw an ice cube tray. Draw 2 rows of ice cubes.
Draw 8 ice cubes in each row. There is 1 layer of ice cubes.

   (Draw your
   picture here.)

   Equation: _____   Volume of ice cube tray: _____

2. Directions: Draw a box of flash cubes for a camera. Draw 2
rows of flashcubes. Draw 3 flashcubes in each row. There is
1 layer of flashcubes.

   Equation: _____   Volume of flashcube box: _____

3. Draw a box of square blocks. There are 2 layers of blocks, so
make the box big enough for 2 layers. Draw 2 rows of blocks.
Draw 3 blocks in each row.

   Equation: _____   Volume of box: _____

4. Draw a box that is big enough for 3 layers. It is a box of number cubes. Show 2 rows of number cubes and 4 number cubes in each row. Write a number on each cube.

Equation: _____   Volume of box: _____

5. Draw a box that is big enough for 3 layers. It is a box of sugar cubes. There are 3 rows of cubes and 3 cubes in each row.

Equation: _____   Volume of box: _____

6. Draw a square box. Write the number of cubes for layers, rows, and cubes in each row. Do not draw the cubes! The height of the box shows how many layers. Write 10 centimeters next to the height of the box. The length and the width of the box show how many rows and how many cubes there are in each row. Write 10 centimeters next to the length of the box. Now write 10 centimeters next to the width of the box.
Drawing:

Equation: _____   Volume of box: _____

B. Now check your drawings with those of a friend. Also check your equations. Does each equation match the box in the

drawing? _____ How does the equation help you find the

volume of the box? _____

Checkup

A. Study each equation. Ring a number or symbol which makes you think of the word in the middle.

1. $3 \times 4 = 12$ factor   **2.** $16 = 12 \times 8$

3. $8 = 2 \times 4$ product   **4.** $5 \times 5 = 25$

5. $9 \times 5 = 45$ equals   **6.** $3 \times 7 = 21$

7. $7 \times 6 = 42$ times   **8.** $8 \times 4 = 32$

B. Match each equation to the property it shows. Write the equation numbers in the blanks next to the properties.

1. $4 \times 5 = 5 \times 4$ _____ Order Property for multiplication

2. $143 \times 0 = 0$

    _____ Zero Property for multiplication

3. $70 \times 1 = 70$

4. $1 \times 36 = 36$ _____ One Property for multiplication

C. Ring the key numbers. Then solve the problems.

1. Jim has 3 coin folders. In one folder he put 16 quarters on each of 4 pages. How many quarters does he have

in all? _____

2. Gloria collected 85 dimes. Her coin folder holds 24 dimes on a page. This folder has 3 pages. How many dimes can she

put in her folder? _____

D. Each picture shows a measurement beside it. Decide whether the picture makes you think of volume or area. Write **volume** or **area** beside the picture.

1. 9 cubic units

_____

2. 9 square units

_____

3. 45 square units

_____

4. 12 cubic units

_____

Division

Look at the picture. Your teacher will tell you what to do.

Vocabulary for Division

A. Work with a friend. Read and talk about the division problems and the division words in the **Division Word Bank.**

$$\overset{5}{3\overline{)15}} \qquad \overset{5\ R3}{5\overline{)28}} \qquad 16 \div 4 = 4$$

DIVISION WORD BANK

dividend	quotient
divisor	remainder

B. Now read and complete each sentence. Use the words in the Division Word Bank.

1. The answer in a division problem is the _____.

2. The _____ is the number that has to be divided.

3. The _____ is the number that you divide by.

4. When you divide a number into equal sets and there is a number left over, that number is the _____.

C. Work with a friend. Your friend should cover the equations left of the dotted line. As you read each equation aloud, your friend writes it in his or her book. For example, you read: "Twelve divided by four is three." Your friend writes: $12 \div 4 = 3$.

1. $9 \div 3 = 3$ | _____

2. $8 \div 2 = 4$ | _____

3. $4 \div 1 = 4$ | _____

4. $6 \div 6 = 1$ | _____

5. $10 \div 5 = 2$ | _____

6. $12 \div 3 = 4$ | _____

7. $14 \div 2 = 7$ | _____

8. $15 \div 3 = 5$ | _____

9. $16 \div 4 = 4$ | _____

10. $20 \div 5 = 4$ | _____

Now you cover the equations. Your friend will read the equations. Listen and write them in your book.

Uncover the equations. Check your work. How many right answers did you have? _____ How many wrong answers did you have? _____ What equations are difficult to understand? _____

Division and Multiplication

The math operations **multiplication** and **division** are related. The **Fact Family** shows how.

FACT FAMILY	SETS	SETS
$4 \times 3 = 12$	3 sets $\times$	4 sets $\times$
$3 \times 4 = 12$	4 per set	3 per set
$12 \div 3 = 4$	= 12	= 12
$12 \div 4 = 3$		

$$12 \div 4 = 3 \text{ sets} \qquad 12 \div 3 = 4 \text{ sets}$$

The Fact Family tells you that finding the quotient is like finding the missing factor in multiplication.

$12 \div 4 = 3$ is like $? \times 4 = 12$.

You can find quotients by thinking of missing factors.

A. Work with a friend. Take turns saying these problems out loud. For problem 1, say: "Twenty-four divided by four is _____."
Then say: "_____ times four equals twenty-four."

1. $24 \div 4 =$ _____   _____ $\times 4 = 24$

2. $42 \div 6 =$ _____   _____ $\times 6 = 42$

3. $35 \div 7 =$ _____   _____ $\times 7 = 35$

4. $36 \div 4 =$ _____   _____ $\times 4 = 36$

B. Now you and your friend find a missing factor to show what happens if you divide any number by itself. Enter the answer for the first problem. Then make up your own.

1. $9 \div 9 =$ _____   ____ $\times 9 = 9$

2. ____ $\div$ ____ $=$ _____   ____ $\times$ ____ $=$ _____

Any number (except zero) divided by itself is equal to _____.

C. With your friend, use missing factors to show what happens if you divide any number by one. For example, what do you get if you divide 12 by 1? Enter the answer for the first problem. Then make up your own.

1. $12 \div 1 =$ _____   ____ $\times 1 = 12$

2. ____ $\div$ ____ $=$ _____   ____ $\times$ ____ $=$ _____

Any number divided by one is equal to _____.

Problem Solving: Division

Plan What to Do

QUESTION
DATA
PLAN
ANSWER
CHECK

Understand the Operation

There are 2 bowls of apples. Each bowl has 4 apples. There are 8 apples in all.

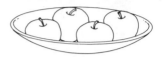   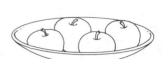

Division can tell how many sets.

$$8 \div 4 = 2$$

There are 2 bowls.

Division can tell how many in each set.

$$8 \div 2 = 4$$

There are 4 in each bowl.

A. Solve the following problems. Use the 5-Point Checklist.

1. There are 15 apples. Put the same number into each of 5 bowls. How many apples are in each bowl?

   Rewrite the question: _____
   Picture:

2. Tania has 36 berries. How many bowls does she need to put 9 in each bowl?

   Rewrite the question: _____
   Picture:

3. 2 workers picked 35 pears. They put the same number into each of 7 bowls. How many pears are in each bowl?

   Rewrite the question: _____
   Picture:

4. 2 cooks used 18 peaches for 6 peach cobblers. About how many peaches did they use for each cobbler?

   Rewrite the question: _____
   Picture:

B. Now check your answers with a friend. Did you rewrite the

questions the same way? _____ Were your answers the

same? _____

Problem Solving: Mixed Practice

QUESTION
DATA
PLAN
ANSWER
CHECK

A. With two friends, read and solve these problems out loud. Explain each step of the 5-Point Checklist. Take turns. When it is your turn to listen, ask a friend to explain steps that you do not understand. Give your friends help on difficult steps by asking questions, but do not give the answers.

1. There were 56 crew members working on Friday. 7 of them were on each flight. How many flights were there? _____

2. A 747 plane has 540 seats. If each row has 6 seats, how many rows are there? _____

3. There were 9 flight crews who reported for work on Monday. There were 9 people in each crew. How many people reported for work? _____

4. One section of the plane has 72 seats. If each row has 8 seats, how many rows are there? _____

5. There were 4 crews with 7 people each and 9 crews with 8 people each. How many people were there? _____

6. The weekend crews had a total of 37 men and 40 women on the 747 flights last week. These crew members worked on 7 flights. How many crew members were on each flight? _____

B. Discuss these questions with your friends: Did you all solve the problems the same way? Did some of you write the problems down? Did others do the problem in their head? Did some estimate the answer? Did you use different ways to check the answer? Which parts of the 5-Point Checklist did you have difficulty with? Write out what you did on the step you had difficulty with:

1. The step I had the most difficulty with was _____

2. The problem number (or numbers) where I had this difficulty was (or were) _____

3. The way I solved this step was to _____

_____

2-Digit Divisors

In many division problems you need to use long division or a calculator to find the answer.

Example: Ana's volleyball team has $64 to buy volleyballs. Each volleyball costs $27. How many volleyballs can the team buy? How much money will they have left? Since each of the volleyballs cost the same amount, we can divide. Study the steps for long division.

Round and Estimate	Multiply	Subtract	Compare
$$\begin{array}{r} 2 \\ 27\overline{)64} \end{array}$$	$$\begin{array}{r} 2 \\ 27\overline{)64} \\ 54 \end{array}$$	$$\begin{array}{r} 2 \\ 27\overline{)64} \\ -54 \\ \hline 10 \end{array}$$	$$\begin{array}{r} 2\ R10 \\ 27\overline{)64} \\ 54 \\ \hline 10 \end{array}$$

The team can buy 2 volleyballs. They will have $10 left over.

A. Study this problem. Write what you do for each step. Round and:

1. _____ 2. _____ 3. _____ 4. _____

$$\begin{array}{r} 4 \\ 12\overline{)49} \end{array}$$
$$\begin{array}{r} 4 \\ 12\overline{)49} \\ 48 \end{array}$$
$$\begin{array}{r} 4 \\ 12\overline{)49} \\ -48 \\ \hline 1 \end{array}$$
$$\begin{array}{r} 4\ R1 \\ 12\overline{)49} \\ 48 \\ \hline 1 \end{array}$$

B. Work these problems with a friend. Explain how you select the quotient and how you figure the remainder.

1. $23\overline{)75}$ 2. $76\overline{)92}$ 3. $48\overline{)97}$ 4. $36\overline{)48}$ 5. $32\overline{)79}$

6. $41\overline{)85}$ 7. $58\overline{)73}$ 8. $44\overline{)92}$ 9. $23\overline{)69}$ 10. $19\overline{)67}$

C. Do you know a way to check your answer? _____ Does your friend know a way? _____ Try ways to check your answer.

Problem Solving: Make a Table

Sample Problem: Juanita is training for a 10-mile track run. She now runs 2 miles every 15 minutes. At this rate, how far will she run in 60 minutes?

A. To solve a problem such as this, you may need to do more than just add, subtract, multiply, or divide. Here is a strategy that might help you.

1. First make a table and write what you know.

Miles	2	4	6	
Minutes	15	30		

2. Now fill in the table to find the answer. When the number reaches 60 minutes, you will have the number of miles in your table. Juanita will run 8 miles in 60 minutes.

B. Solve these problems out loud with two friends. Explain each step as you fill in the table. Take turns. When it is your turn to listen, ask a friend to explain if you do not understand. Give your friends help on difficult parts by asking questions, but do not give the answers.

1. The buttons you want come in packages of 6 buttons for 80¢ a package. You need 32 buttons. You have a five dollar bill. Do you have enough money? Complete the table to decide.

Buttons	6	12	18	24	30	36
Cost per package	80¢	$1.60				

2. It takes 2 minutes to play 3 video games. How long does it take to play 21 games? Complete the table.

Games							
Minutes							

C. Discuss with your friends which parts were hard to explain and which parts were easy. Check with another team if you cannot get the answers.

Write Your Own Problems

Now it is your turn to write division problems. Follow these steps:

1. Make a Data Table.
2. Choose the data.
3. Organize your ideas.
4. Write a word problem.
5. Read your problem to two friends.
6. Solve each other's problems.
7. Check your answers.

A. Work with one or more friends to count things and people in your classroom. Record the data on the **Classroom Data Table.**

CLASSROOM DATA TABLE		
Numbers of Things	**Numbers of People**	**Numbers of Animals**
_____ desks	_____ teachers	_____ hamsters
_____ chairs	_____ girls	_____ fish
_____ tables	_____ boys	_____ birds
_____ bookshelves	_____ others	_____ others
_____ library books		
_____ pieces of chalk		
_____ pencils		

B. Now use the Classroom Data Table to write a division problem.

1. Choose a number that tells how many things.

2. Choose another number that tells how many people or animals.

3. Write a division equation that shows how the number of things can be divided equally between the number of people or animals.

4. Write a story about the numbers in the equation.

For example, if there are 24 pieces of chalk in your classroom and 2 teachers, you can write a division problem like this:

Division equation: $24 \div 2 = 12$

Story: There are 2 teachers in my class and 24 pieces of chalk. The teachers want to divide the chalk equally.

Question: How many pieces of chalk does each teacher get?

C. Use the data in the Classroom Data Table to write division problems. Work by yourself.

Problem 1

Division fact: _____

Story: _____

_____

Question: _____

Problem 2

Division fact: _____

Story: _____

_____

Question: _____

Problem 3

Division fact: _____

Story: _____

_____

Question: _____

Problem 4

Division fact: _____

Story: _____

_____

Question: _____

D. Sit with two friends. Take turns reading your problems and solving them. Use this space to solve your friends' problems.

E. Check your answers with your friends. Write a sentence that

tells what was difficult. _____

Estimation: Mental Math

You can buy 4 packages of colored pencils for $7.56. About how many dollars does 1 package of colored pencils cost?

Since you want an answer that is only **close** to the exact answer, you **estimate** by rounding and dividing in your head.

$7.56 ÷ 4   (THINK) $8 ÷ 4 = $2

The cost of 1 package is about $2.

A. Complete these other examples of quotients with money.

$9.10 ÷ 3 = ? $11.95 ÷ 2 = ?

  $9 ÷ 3 = _____ $12 ÷ 2 = _____

B. Estimate these quotients. Take turns with a friend. Say each equation aloud. Round to the nearest dollar. Write the new equation with the rounded amounts and divide.

1. $4.17 ÷ 4 = n **2.** $6.05 ÷ 3 = n

_____ ÷ _____ = _____   _____ ÷ _____ = _____

3. $5.75 ÷ 2 = n **4.** $3.89 ÷ 2 = n

_____ ÷ _____ = _____   _____ ÷ _____ = _____

5. $7.15 ÷ 7 = n **6.** $9.68 ÷ 5 = n

_____ ÷ _____ = _____   _____ ÷ _____ = _____

C. Now do both the rounding and dividing in your head. Just write the answers.

1. $11.79 ÷ 4 = _____   **2.** $9.11 ÷ 3 = _____

3. $14.99 ÷ 3 = _____   **4.** $13.95 ÷ 2 = _____

5. $24.05 ÷ 4 = _____   **6.** $55.55 ÷ 8 = _____

D. Discuss your answers. Which items did you have difficulty

with? _____ Which ones were easy? _____

What made the hard ones difficult for you? _____

Problem Solving with Estimation

After you solve a problem you should check to see if your answer is reasonable. You should also check to see if your answer is correct.

   You can use estimating to see if your answer is reasonable. To see if your answer is correct in a division problem you can multiply the quotient by the divisor and then add the remainder.

QUESTION
DATA
PLAN
ANSWER
CHECK

Sample:

$$\frac{\text{quotient}}{\text{divisor)dividend}}$$

$$\begin{array}{r} 9 \text{ R2} \\ 4\overline{)38} \\ -36 \\ \hline 2 \end{array}$$

$$\begin{array}{r} 9 \\ \times\ 4 \\ \hline 36 \\ +\ 2 \\ \hline 38 \end{array}$$

A. Work with a friend. For each problem decide if the answer given is reasonable. If the answer is reasonable, check to see if it is correct. Give the correct answer when the given one is wrong.

   1. 87 students went on a picnic in buses. Each bus holds 28 students. How many buses do they need? Bob said they need at least 4 buses. What do you think about Bob's

   answer? _____

   _____

   What would you give for the answer? _____

   2. 95 people at the picnic will want a hot dog. The hot dogs come 8 to a package. How many packages are needed? Gladys said they need 24 packages. How many packages do

   you think are needed? _____

   Show your friend how you found your answer. Check to see if it is correct.

B. Check each quotient and remainder. Place an X beside each incorrect answer. Work with a friend. Take turns. Explain your method for each one you check.

   1. $\dfrac{4 \text{ R6}}{12\overline{)45}}$ 2. $\dfrac{4 \text{ R2}}{22\overline{)90}}$ 3. $\dfrac{16 \text{ R0}}{8\overline{)128}}$

   4. $\dfrac{6 \text{ R3}}{9\overline{)49}}$ 5. $\dfrac{13 \text{ R1}}{35\overline{)456}}$ 6. $\dfrac{9 \text{ R10}}{11\overline{)134}}$

C. Compare your answers with a friend's. Do they agree? _____

   Which kinds of problems do you prefer? Why? _____

   _____

   _____

Finding Averages

Paco, Omar, and Mary Lou wanted to find out their average height. They measured each other's heights with a yardstick. Paco was 61 inches tall. Omar was 57 inches tall. Mary Lou was 59 inches tall.

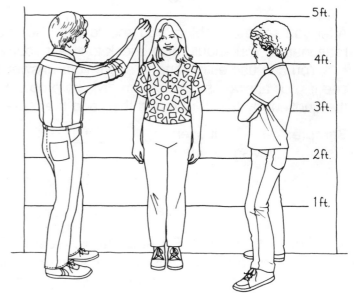

To find the average of these 3 numbers, they added the numbers and then divided by 3. Their average height is 59 inches.

```
  61        59
  57     3)177
+ 59       15
-----      27
 177       27
```

When we want to find the average of a list of numbers, we add the numbers and divide by the number of numbers we added.

A. Measure the heights of 3 classmates in inches. Write their heights:

Name _____   Height _____

Name _____   Height _____

Name _____   Height _____

Add the 3 heights: _____   Divide the total by 3: _____

What is the average height of these 3 people? _____

B. Now choose 4 classmates. Measure their heights in inches. Then find the average height of these 4 people.

Name _____   Height _____

Name _____   Height _____

Name _____   Height _____

Name _____   Height _____

Use this space for calculating:

What is the average height of these 4 people? _____

C. Now choose 5 classmates. Measure their heights in inches. Then find the average height of these 5 people.

Name _____ Height _____

Name _____ Height _____

Name _____ Height _____

Name _____ Height _____

Name _____ Height _____

Use this space for calculating:

What is the average height of these 5 people? _____

D. With your friend, measure the lengths of these classroom objects. Find the average length for each object.

PENCILS	PENS	PIECES OF CHALK
Pencil 1 _____	Pen 1 _____	Piece 1 _____
Pencil 2 _____	Pen 2 _____	Piece 2 _____
Pencil 3 _____	Pen 3 _____	Piece 3 _____
Pencil 4 _____	Pen 4 _____	
Sum _____	Sum _____	Sum _____
Average _____	Average _____	Average _____

E. When you divide a sum to find the average, how do you know what number to use as the divisor? (Answer in a complete

sentence.) _____

_____

_____

Read your sentence to a friend. Listen while your friend reads

his or her sentence to you. Are the sentences alike? _____

F. Work with a friend to check your addition and division calculations. Then work together to check your friend's addition and division calculations. (You may use a calculator.)

Checkup

Work by yourself.

A. Put an X on the divisor.
Ring the quotient.
Underline the dividend.

$$\begin{array}{r} 3\ \text{R10} \\ 22\overline{)76} \\ \underline{66} \\ 10 \end{array}$$

B. Write and solve a division equation for each problem your teacher reads.

1. _____

2. _____

C. Read the problem. Rewrite the question. Draw a picture. Solve the problem.

María has 18 beads. She puts 6 beads on a string. How many strings does she have?

_____

_____

D. Solve these problems.

1. A 747 plane has 540 seats. If each row has 9 seats, how many seats are there

in all? _____

2. Dave's soccer team has $76 to buy soccer balls. If each ball costs $28, how

many balls can they buy? _____
How much money will be left

over? _____

E. Check each answer. Mark **correct** or **incorrect**. If an answer is incorrect, give the correct one.

1. $\begin{array}{r} 7\ \text{R2} \\ 9\overline{)68} \end{array}$

_____ correct

_____ incorrect

_____ my answer

2. $\begin{array}{r} 4\ \text{R2} \\ 12\overline{)50} \end{array}$

_____ correct

_____ incorrect

_____ my answer

Four Operations

Look at the picture. Your teacher will tell you what to do.

Vocabulary Review

A. Write the math sign (+, −, ×, ÷) that shows which operation each word or phrase tells you to do. Some words and phrases can be used for more than one operation. The first one is done for you.

1. less ___−___
2. times _____
3. in all _____
4. find the answer _____
5. quotient _____
6. dividend _____
7. total _____
8. altogether _____
9. how many? _____
10. greater than _____
11. minus _____
12. addend _____
13. divisor _____
14. how many in each? _____
15. sum _____
16. how much? _____
17. more than _____
18. factor _____
19. find the average _____
20. difference _____
21. solve the problem _____
22. product _____
23. remainder _____
24. less than _____

B. Check your work with a friend. Correct your wrong answers on the lines below. Write the word or phrase. Then write the correct math sign or signs next to it.

_____   _____

_____   _____

C. For each sign, write the English word next to it. Choose from this list: add, subtract, multiply, divide.

+ _____   × _____

− _____   ÷ _____

D. Write a word or phrase that tells you to do **two** operations:

_____

Reading Ordinal Numbers

We use **ordinal numbers** to tell the position of something. Look at these examples of ordinal numbers:

María is **first** in line.
Bob made the **second** goal.
The **fifth** book on the shelf is a dictionary.

A. Look at the ordinal numbers and names as your teacher reads them:

1st	first	11th	eleventh	21st	twenty-first
2nd	second	12th	twelfth	22nd	twenty-second
3rd	third	13th	thirteenth	23rd	twenty-third
4th	fourth	14th	fourteenth	24th	twenty-fourth
5th	fifth	15th	fifteenth		
6th	sixth	16th	sixteenth	30th	thirtieth
7th	seventh	17th	seventeenth	31st	thirty-first
8th	eighth	18th	eighteenth	32nd	thirty-second
9th	ninth	19th	nineteenth	33rd	thirty-third
10th	tenth	20th	twentieth	34th	thirty-fourth

B. Now write.

1. Write 3 ordinal numbers that end in **-st.** _____

2. Write 3 ordinal numbers that end in **-nd.** _____

3. Write 3 ordinal numbers that end in **-rd.** _____

4. What letters do the other ordinal numbers end in? _____

C. Work with a friend. Look at page 67. People are waiting for some of the rides. Take turns asking and answering questions about the people in the various lines. Point to the person you are talking about when you give the answer.

1. Who is first in the roller coaster line?

2. Who is third in the merry-go-round line?

3. Who is sixth in the merry-go-round line?

4. Who is tenth in the roller coaster line?

5. Who is fifth in the merry-go-round line?

6. Who is eighth in the merry-go-round line?

7. Who is second in the roller coaster line?

8. Who is ninth in the roller coaster line?

Problem Solving: Choose the Operations

QUESTION
DATA
PLAN
ANSWER
CHECK

Remember:

PUT TOGETHER ADD +	TAKE AWAY SUBTRACT −	COMPARE SUBTRACT −
PUT TOGETHER EQUAL SETS MULTIPLY ×	HOW MANY SETS? DIVIDE ÷	HOW MANY IN EACH SET? DIVIDE ÷

A. Work with a friend. Take turns reading each problem aloud. **Choose the operations** together. Decide whether you have to **add, subtract, multiply,** or **divide.** You may have to do more than one operation in a problem!

    Write the names of the operations (addition, subtraction, multiplication, division) on the line after each problem. *Do not solve the problems yet.*

1. Miguel bought nine 40¢ stamps and three 5¢ stamps at the post office. How much did he pay in all?

Operation(s): _____

Answer _____

2. On Monday Bob mailed a package. It cost $1.36. On Friday he mailed another package, and paid $2.12. How much less did he pay on Monday?

Operation(s): _____

Answer: _____

3. A clerk at the post office sold 7 books of stamps. There are 30 stamps in each book. How many stamps did the clerk sell?

Operation(s): _____

Answer: _____

4. Nu Trinh put the same number of stamps on 6 packages. She used 18 stamps. How many stamps are on each package?

Operation(s): _____

Answer: _____

5. Elena mailed 3 packages. One package cost $2.76 to mail. The second cost $0.82. The third cost $1.55. How much did Elena pay?

Operation(s): _____

Answer: _____

6. Chui bought seven 20¢ stamps. How much did he pay? He gave the clerk $5. How much money did he get back?

Operation(s): _____

Answer: _____

_____

B. Now work by yourself to solve each problem in Section A. Use the space at the right of each problem for calculations. Write the answer in a complete sentence on the answer line.

C. Work with three or four friends. Check your work together. Did everyone in the group choose the same

operations? _____ Which problems were

easy? _____ Which were difficult? _____

Using the 5-Point Checklist

Many people like to collect things. Some people collect old coins. These are some old American coins that people collect:

Indian Head Penny

Lincoln Penny

Large Cent

Liberty Dime

A. Do you remember the 5-Point Checklist? Fill in the missing words.

1. Understand the _____.
2. Find the needed data.

3. _____ what to do.
4. Find the answer.

5. _____ back.

B. Work in groups of four or five. Take turns reading each problem aloud. Do the first 3 steps of the 5-Point Checklist. Rewrite the question, find the data, and plan what to do. In each rewritten question, leave a blank for the answer.

   1. Leila has a 1901 Indian Head Penny. She also has a Lincoln Penny. The Lincoln Penny was made 28 years after the Indian Head Penny. When was the Lincoln Penny made?

   a. Understand the Question

   Rewrite the question: _____
   b. Find the Needed Data

   Data needed: _____
   c. Plan What to Do

   Operation? _____

   2. Franco has one page of Lincoln Pennies. The page has 7 rows with 5 pennies in each row. How many pennies does Franco have?

   a. Understand the Question

   Rewrite the question: _____
   b. Find the Needed Data

   Data needed: _____

c. Plan What to Do

Operation? _____

3. Carmen has 24 coins to put in her book. She puts 4 coins in each row. How many rows does she have?

   a. Understand the Question

   Rewrite the question: _____

   b. Find the Needed Data

   Data needed: _____

   c. Plan What to Do

   Operation? _____

4. May Chu has 8 Indian Head Pennies and 12 Lincoln Head Pennies. She put them into 5 equal rows. How many did she put in each row?

   a. Understand the Question

   Rewrite the question: _____

   b. Find the Needed Data

   Data needed: _____

   c. Plan What to Do

   Operation? _____

5. Ho sold 8 Large Cents for $6 each. He also sold a Liberty Dime for $7. How much money did he get?

   a. Understand the Question

   Rewrite the question: _____

   b. Find the Needed Data

   Data needed: _____

   c. Plan What to Do

   Operation? _____

B. Work by yourself. Go back and do step 4. Find the answer for each problem in Section A. Write the answer in the blank where you rewrote the question.

C. Work with a friend. Do step 5 together. Check back. (You may use a calculator.) Number right: _____ Which problems were difficult? _____

Why? _____

Write Your Own Problems

Carnival Time

Miguel's class went to a carnival with lots of games and rides.
Each person in the class got to choose what activity to do.

A. Use the chart at the right to make up
your own word problems. Use two
operations in a problem if possible.
Choose *all four* operations in the different
problems you do. One sample problem,
with two operations, is done for you.
Work by yourself.

School Carnival	
Activity	**Cost**
Ring Toss	25¢
Baseball Throw	30¢
Ferris Wheel	65¢
Bumper Cars	50¢
Spider Ride	45¢
Water Slide	60¢
Carousel	35¢
Roller Coaster	75¢

Problem 1

Math facts: 25¢ × 4 persons and 35¢ × 8 persons.

Story: Miguel and 3 of his friends did the Ring Toss. 8 of the other children went on the Carousel.

Question: What was the total amount both groups of children spent?

Problem 2

Math facts: _____

Story: _____

_____

Question: _____

Problem 3

Math facts: _____

Story: _____

_____

Question: _____

Problem 4

Math facts: _____

Story: _____

_____

Question: _____

Problem 5

Math facts: _____

Story: _____

_____

Question: _____

B. Now sit with a friend. Show your problems to your friend. Tell your friend to use the 5-Point Checklist. Also, tell your friend to say each step out loud. When it is your turn to solve a problem, make a table if you have difficulty.

Use this space to solve your friend's problems.

```
QUESTION
DATA
PLAN
ANSWER
CHECK
```

C. Now check your answers with two other friends. (You may use a calculator.) Look at the problems that were difficult. Write a sentence that tells what was difficult.

_____

Space Figures

A. Space figures are like many objects in the world around you. Work with a friend. Tell what space figure each object suggests. Write the letter of the object next to the name of the figure.

1. Sphere _____

_____

2. Cylinder _____

_____

3. Rectangular prism _____

_____

4. Cone _____

_____

5. Cube _____

_____

6. Pyramid _____

_____

A

B

C

D

E

F

G

B. Look around your classroom. Find other objects to fit the figures listed above. Write their names in the spaces provided in Section A.

C. Choose one of the space figures. Write two or three complete sentences to tell what objects around you at school, or at home,

remind you of that figure. Explain why for each one. _____

_____

_____

D. Work with a friend. Take turns reading each other's sentences. When your friend reads your sentences, do they sound the way

you intended? _____

E. On a separate sheet, draw space figures from memory. Work with a friend, who will name the figures to you: sphere, cube, rectangular prism, pyramid, cone, cylinder. Then you name them while your friend draws them from memory. Compare your answers afterward.

Plane Figures

Space figures fill up a space. Plane figures lie flat.

Squares and Rectangles

The face of a cube is a square.

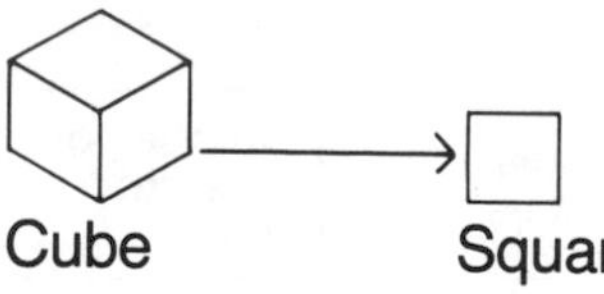

Cube → Square

The face of a rectangular prism is a rectangle.

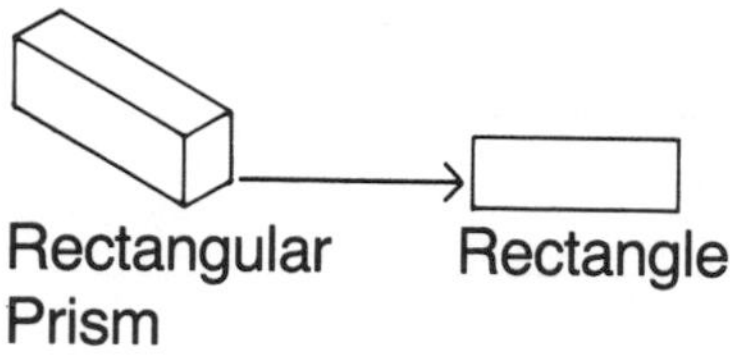

Rectangular Prism → Rectangle

A. Name some surfaces around the room that make you think of a square or rectangle.

1. Square

_____

_____

_____

2. Rectangle

_____

_____

_____

Show your list to a friend. Did you list the same

things? _____ Do you agree with your friend's

list? _____ Does your friend agree with your list? _____

Circles and Triangles

A face of a cylinder has a circle shape.

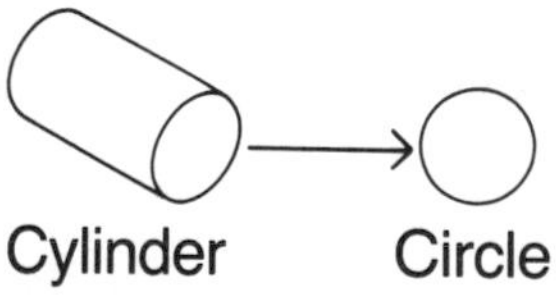

Cylinder → Circle

Think of cutting off a corner of a cube to get the shape of a triangle.

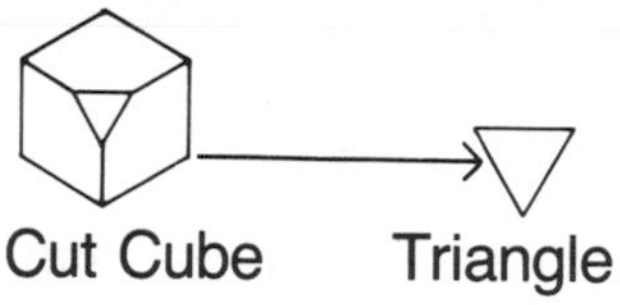

Cut Cube → Triangle

B. List some things that make you think of a circle or triangle.

1. Circle _____

_____

_____

2. Triangle _____

_____

_____

Compare your list with a friend's. Did your friend think of things

you missed? _____ If so, add them to your list.

Faces, Vertices, and Edges

A. Work with a friend. Complete the table. Use objects when you need to.

Name	Number of Flat Faces	Number of Vertices	Number of Straight Edges	Number of Curved Edges	Number of Curved Faces
1. Cube		8	12		
2. Rectangular Prism				0	
3. Sphere					1
4. Cone				1	
5. Cylinder	2	0	0		1
6. Pyramid	5	5			

B. Work with a friend. Complete each sentence. Discuss with your friend what you should write in the blanks.

1. A plane figure that has 3 sides and 3 corners is

    a _____.

2. A box of breakfast cereal makes us think of

    a _____.

3. A plane figure that has two pairs of equal length sides

    is a _____.

4. A _____ has no faces, no vertices, and no edges.

5. A _____ has equal faces, 8 vertices, and 12 straight edges.

C. Look at this list of names of plane and space figures. Underline the words you used in Section B. Compare your list with a

friend's. Do they agree? _____ Do some sentences have

more than one correct answer? _____

circle   triangle   square   rectangle   pyramid
cone   sphere   cube   cylinder   rectangular prism

Checkup

Work alone.

A. Write the math sign (+, −, ×, ÷) that shows which operation each phrase or word makes you think about.

1. minus _____   **2.** addend _____   **3.** more than _____

4. times _____   **5.** dividend _____   **6.** total _____

7. sum _____   **8.** quotient _____   **9.** divisor _____

B. Match the ordinal numbers with the correct words. Write each ordinal in the blank where it belongs.

1. 3rd _____ twenty-first

2. 21st _____ fiftieth

3. 15th _____ fifteenth

4. 50th _____ third

C. Choose the operation(s) you want to use to solve each problem: add, subtract, multiply, or divide. Then find the answer.

1. Carmen paid $0.78 to mail an envelope and $1.62 for a package. How much more did she pay for the package?

Operation(s) _____

Answer _____

2. Ali bought five 22¢ stamps and ten 14¢ stamps. What was the total cost of the stamps?

Operation(s) _____

Answer _____

D. **1.** Name 3 space figures. Draw the figures in the space provided.

_____

2. Name 3 plane figures. Draw the figures in the space provided.

_____

Fractions

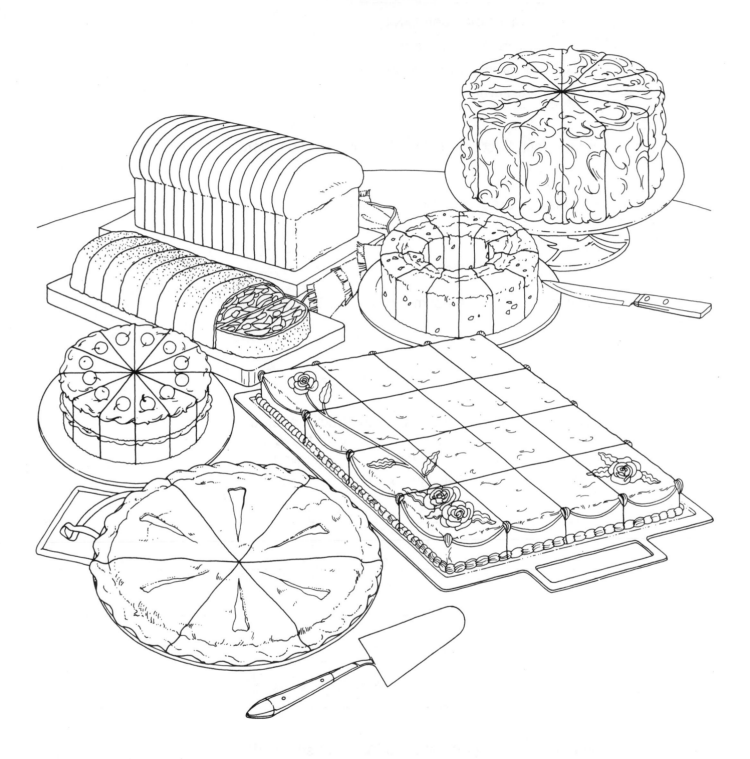

Look at the picture. Your teacher will tell you what to do.

Vocabulary for Fractions

A. Work with a friend. Read aloud and talk about the fraction words in the **Fraction Word Bank.**

> **FRACTION WORD BANK**
>
> numerator, denominator,
>
> equivalent fractions, fraction,
>
> mixed numeral

B. Write the word for each fraction problem. Use the Fraction Word Bank.

1. $\frac{3}{5}$ a number that is part of a whole: _____

2. $\frac{2}{3}$ ← the top part of a fraction: _____
   ← the bottom part of a fraction: _____

3. $\frac{1}{3} = \frac{2}{6}$ fractions that have the same value: _____

   _____

4. $2\frac{1}{2}$ a whole number and a fraction together: _____

C. Work with a friend. Your friend should cover the fractions left of the dotted line. As you read the fractions aloud, your friend writes them in his or her book. For example, for $\frac{3}{4}$ you would read: "Three fourths. Three of four equal parts."

1. $\frac{5}{6}$ _____

2. $\frac{2}{3}$ _____

3. $\frac{7}{8}$ _____

4. $\frac{1}{2}$ _____

5. $\frac{3}{4}$ _____

D. Ask your friend to make up some fractions. As he or she reads them, you write them down.

Equivalent Fractions

$\frac{1}{2}$ of the strip is shaded ⇨ [strip diagram]

$\frac{3}{6}$ of the strip is shaded ⇨ [strip diagram]

Each strip is shaded the same amount.

$$\frac{1}{2} = \frac{3}{6}$$

These are **equivalent fractions,** because they name the same amount.

A. Listen to your teacher for directions. You will fold strips to show one-half, two-fourths, and four-eighths.

B. Do the following.

   1. Complete this equation showing 2 equivalent fractions. (Use the strips.) $\frac{1}{2} = \frac{}{4}$

   2. Write another equation showing 2 equivalent fractions that

    match two of the strips: _____
               Check your answers with a friend.

C. Work with your friend. Take turns using each picture to name a fraction that is equivalent to $\frac{1}{2}$. Tell your friend how to work the problem. Say the answer. Then write it in the space.

1.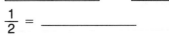
$\frac{1}{2} =$ _____

2.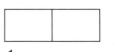
$\frac{1}{2} =$ _____

3.
$\frac{1}{2} =$ _____

4.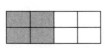
$\frac{1}{2} =$ _____

D. What is the missing fraction? Tell your friend how to work each problem. Say the answer. Then write it in the space.

1.
$\frac{3}{4} =$ _____

2.
$\frac{1}{3} =$ _____

3.
$\frac{2}{4} =$ _____

Write a sentence that tells which ones were hard and why.

_____

Lowest-Terms Fractions

The store had only 30 cans of apple juice left.
Luisa bought 12 of them.

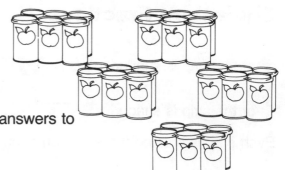

A. Form a group of four or five students. Figure out the answers to
these questions. Your teacher will help if necessary.

   1. What fraction of the apple juice cans did Luisa

   buy? _____

   2. There are 6 cans in each case. So what fraction of the cases

   did Luisa buy? _____ Tell why the fraction of apple juice
   cans is the same as the fraction of cases. $\frac{12}{30} = \frac{2}{5}$

The problem you just worked shows how to use division to "reduce"
a fraction to an equivalent fraction. An **equivalent fraction** is one
that is the same amount.

To **reduce a fraction,** divide both the numerator and the denominator by a whole number greater than 1.	A fraction is in **lowest terms** when it cannot be reduced.

Other Examples

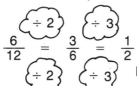

$$\frac{6}{12} = \frac{3}{6} = \frac{1}{2}$$ Lowest Terms

$$\frac{8}{12} = \frac{2}{3}$$ Lowest Terms

$$\frac{6}{10} = \frac{3}{5}$$ Lowest Terms

B. Work with the same small group. Take turns. Tell your friends
how to reduce each fraction to lowest terms. Say the answer.
Then write it in the space.

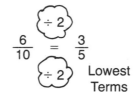

   1. $\frac{6}{8} = \frac{3}{4}$ **2.** $\frac{6}{15} =$ ____   **3.** $\frac{12}{16} =$ ____

   4. $\frac{4}{6} =$ ____   **5.** $\frac{2}{10} =$ ____   **6.** $\frac{10}{16} =$ ____   **7.** $\frac{9}{12} =$ ____

   8. $\frac{3}{9} =$ ____   **9.** $\frac{8}{20} =$ ____   **10.** $\frac{4}{16} =$ ____   **11.** $\frac{10}{14} =$ ____

C. Tell your friends which ones were easy and which ones were
hard. Write a sentence that describes what made them easy
or hard.

Comparing Fractions

A. Use strips of yarn to help you compare fractions.

Halves

Thirds

Fourths

Fifths

Tenths

Examples $\frac{1}{3}$ is greater than $\frac{1}{4}$ $\longrightarrow$ $\frac{1}{3} > \frac{1}{4}$

$\frac{1}{4}$ is greater than $\frac{2}{5}$ $\longrightarrow$ $\frac{1}{4} > \frac{2}{5}$

$\frac{1}{2}$ is equal to $\frac{2}{4}$ $\longrightarrow$ $\frac{1}{2} = \frac{2}{4}$

B. Play tic-tac-toe with a friend. You get to enter an X or an O *only* if you can tell your friend *how* to answer the question. You give up your turn if you cannot tell **how** you got the answer. Otherwise, the rules of tic-tac-toe are the same. Use the fractions in Section C.

Rules: Take turns marking an X or an O in the boxes. The first player to place 3 X's or O's in a straight line wins.

C. Write the sign >, <, or = in each circle. Take turns and play tic-tac-toe.

1. $\frac{1}{2} \bigcirc \frac{1}{3}$ 2. $\frac{1}{5} \bigcirc \frac{1}{3}$ 3. $\frac{1}{10} \bigcirc \frac{1}{5}$ 4. $\frac{1}{4} \bigcirc \frac{1}{5}$

5. $\frac{2}{5} \bigcirc \frac{5}{10}$ 6. $\frac{1}{3} \bigcirc \frac{1}{10}$ 7. $\frac{2}{5} \bigcirc \frac{1}{2}$ 8. $\frac{1}{5} \bigcirc \frac{3}{10}$

9. $\frac{1}{5} \bigcirc \frac{2}{10}$ 10. $\frac{3}{4} \bigcirc \frac{2}{3}$ 11. $\frac{3}{5} \bigcirc \frac{6}{10}$ 12. $\frac{7}{10} \bigcirc \frac{4}{5}$

13. $\frac{5}{10} \bigcirc \frac{4}{5}$ 14. $\frac{8}{10} \bigcirc \frac{4}{5}$ 15. $\frac{1}{2} \bigcirc \frac{2}{3}$ 16. $\frac{2}{3} \bigcirc \frac{2}{5}$

17. $\frac{4}{10} \bigcirc \frac{3}{5}$ 18. $\frac{2}{5} \bigcirc \frac{1}{4}$ 19. $\frac{1}{5} \bigcirc \frac{1}{4}$ 20. $\frac{9}{10} \bigcirc \frac{4}{5}$

Practice in Comparing Fractions

A. For each pie, mark off the amount shown by the fraction. The first one is done for you. Now you do the rest. **Hint:** In the second one, start by dividing the pie into six equal parts. Then shade the part shown by the fraction.

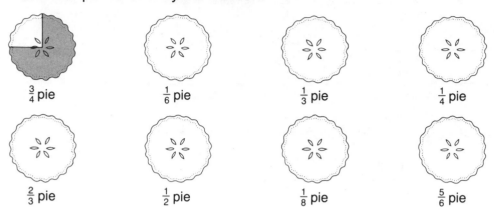

| $\frac{3}{4}$ pie | $\frac{1}{6}$ pie | $\frac{1}{3}$ pie | $\frac{1}{4}$ pie |

| $\frac{2}{3}$ pie | $\frac{1}{2}$ pie | $\frac{1}{8}$ pie | $\frac{5}{6}$ pie |

Now compare the amount you shaded with the amount shaded by a friend. Was it the same for each pie? If you did not agree, check with another team until you all agree.

B. Work with a friend. In the first blank, write a fraction that matches each statement. Use the fractions in Section A for your choices. Take turns.

1. Reynaldo had more than $\frac{1}{6}$ pie but less than $\frac{1}{3}$ pie.

   ___ _____

2. Keiko had less than $\frac{1}{2}$ pie but more than $\frac{1}{4}$ pie.

   ___ _____

3. Sean had more than $\frac{1}{2}$ pie but less than $\frac{3}{4}$ pie.

   ___ _____

4. Marcello had more than $\frac{1}{3}$ pie but less than $\frac{2}{3}$ pie.

   ___ _____

5. Manuel had less than $\frac{5}{6}$ pie but more than $\frac{2}{3}$ pie.

   ___ _____

6. Celina had less than a whole pie but more than $\frac{3}{4}$ pie.

   ___ _____

7. Lin had less than $\frac{1}{4}$ pie but more than $\frac{1}{8}$ pie.

   ___ _____

8. Tran had a small piece that was less than $\frac{1}{6}$ pie.

   ___ _____

C. In the second blank, write two inequality statements for each problem in Section B. Take turns reading the inequalities to your friend. Remember: > means "greater than" and < means "less than."

Adding Fractions

Tim, Janell, and Rosa were all asked this question: How far along the road is it from Ramón's house to Tsau Lin's?

Tim counted the $\frac{1}{10}$ marks. Janell added $\frac{3}{10}$ and $\frac{4}{10}$. Rosa added $\frac{1}{2} + \frac{2}{10}$. Each of them got $\frac{7}{10}$ for their answer.

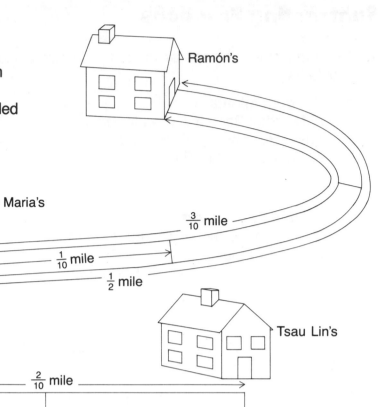

Janell's Work

$$\begin{array}{r} \frac{3}{10} \\ + \frac{4}{10} \\ \hline \frac{7}{10} \end{array}$$

Rosa's Work

$$\begin{array}{r} \frac{1}{2} = \frac{5}{10} \\ + \frac{2}{10} = \frac{2}{10} \\ \hline \frac{7}{10} \end{array}$$

A. Work with a friend. Explain why Rosa needed an extra step. Then solve these problems. Read each problem aloud. Take turns.

1. $\frac{4}{8} + \frac{1}{8} =$ _____

2. $\frac{3}{5} + \frac{1}{5} =$ _____

3. $\frac{4}{10} + \frac{1}{10} =$ _____

4. $\frac{2}{8} + \frac{4}{8} =$ _____

5. $\begin{array}{r} \frac{1}{3} = \\ + \frac{1}{6} = \\ \hline \end{array}$

6. $\begin{array}{r} \frac{1}{4} = \\ + \frac{1}{2} = \\ \hline \end{array}$

7. $\begin{array}{r} \frac{2}{3} \\ + \frac{1}{3} \\ \hline \end{array}$

8. $\begin{array}{r} \frac{3}{8} \\ + \frac{1}{4} \\ \hline \end{array}$

9. $\begin{array}{r} \frac{1}{3} \\ + \frac{2}{9} \\ \hline \end{array}$

B. Write a sentence or two telling why some problems need an extra step. Use the term **common denominator** in your writing.

_____

_____

Subtracting Fractions

A. Use the picture on page 87 to help you solve these subtraction problems.

1. Like denominators
On Monday Ramón walked to María's house. How much farther must he walk to get to Tsau Lin's house?

$$\begin{array}{r} \frac{7}{10} \\ -\frac{3}{10} \\ \hline \frac{4}{10} \end{array}$$

_____

2. Unlike denominators
On Tuesday Ramón walked to Paolo's house. How much farther must he walk to get to Tsau Lin's from Paolo's?

$$\begin{array}{r} \frac{7}{10} = \frac{7}{10} \\ -\frac{1}{2} = \frac{5}{10} \\ \hline \frac{2}{10} \end{array}$$

_____

B. Work with a friend. Explain the difference between Problem 1 and Problem 2 in Section A.

How are Problem 1 and Problem 2 alike? _____

How are they different? _____

C. Solve each problem. Ring the problems that subtract fractions with unlike denominators. Tell your friend how you subtract. Take turns.

1. $\frac{9}{10} - \frac{3}{10} =$ _____

2. $\frac{3}{4} - \frac{1}{4} =$ _____

3. $\frac{3}{7} - \frac{1}{7} =$ _____

4. $\begin{array}{r} \frac{7}{8} = \\ -\frac{1}{4} = \\ \hline \end{array}$

5. $\begin{array}{r} \frac{3}{10} \\ -\frac{1}{10} \\ \hline \end{array}$

6. $\begin{array}{r} \frac{5}{8} = \\ -\frac{1}{2} = \\ \hline \end{array}$

7. $\begin{array}{r} \frac{4}{5} \\ -\frac{2}{5} \\ \hline \end{array}$

8. $\begin{array}{r} \frac{2}{3} = \\ -\frac{1}{6} = \\ \hline \end{array}$

9. $\begin{array}{r} \frac{1}{2} = \\ -\frac{1}{4} = \\ \hline \end{array}$

D. Check your answers with two other friends. Ask questions about any answers you disagree with. Which problems were difficult for you? Write a sentence telling what was difficult.

Problem Solving with Fractions

The 5-Point Checklist is useful for solving
these problems with fractions.

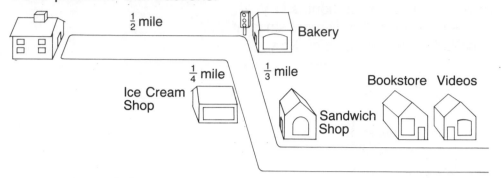

A. Work with a friend. Read each problem aloud. Underline the
question. Ring the key words. Choose an operation. Solve for
the answer. Check to see that your answer is reasonable.

1. Tsau Lin walked $\frac{1}{2}$ mile to the stoplight. Then she walked $\frac{1}{4}$
   mile to the Ice Cream Shop. How far did she walk in all?

   Rewrite the question _____
   Fill in the answer.

2. The Sandwich Shop is $\frac{1}{3}$ mile from the Bakery. Videos is $\frac{5}{6}$
   mile from the Bakery. What is the difference in these
   distances?

   Rewrite the question _____
   Fill in the answer.

3. The Bookstore is $\frac{3}{4}$ mile from the Bakery. Videos is $\frac{5}{6}$ mile
   from the Bakery.

   Which is farther? _____

   How much farther? _____

4. Tsau Lin walked $\frac{1}{2}$ mile to the stoplight. Then she walked $\frac{1}{3}$
   mile farther to the Sandwich Shop. How far did she walk in

   all? _____

5. At the Sandwich Shop she bought $\frac{3}{4}$ quart of potato salad
   and $\frac{1}{2}$ quart of macaroni salad. How much more potato salad

   did she buy than macaroni salad? _____

   How many quarts of salad did she buy in all? _____

B. Check your answers with another friend. Do you disagree on

any answers? _____ Explain to your friend how you found
your answers. Listen as your friend explains his or her answers.
Go back and correct any answers you would like to change.

Estimating with Fractions

The circles below tell about the weather in the towns of Mountain View and Lakeside. Each full circle stands for 12 months (1 year). The parts of each circle show, with fractions, the different kinds of weather that Mountain View and Lakeside had for a whole year.

Lakeside   Mountain View

A. Work with a friend. Take turns reading each problem. Look at the circles to **estimate** the answers to the questions. Ring the best answer.

1. There is snow in Mountain View for about what part of the year?

    a. $\frac{1}{3}$ of the year   **b.** $\frac{1}{2}$ of the year   **c.** $\frac{1}{4}$ of the year

2. There is sunshine in Lakeside for about what part of the year?

    a. $\frac{1}{4}$ of the year   **b.** $\frac{1}{3}$ of the year   **c.** $\frac{1}{2}$ of the year

3. For about what part of each year is there fog in Mountain View?

    a. $\frac{1}{3}$ of the year   **b.** $\frac{1}{6}$ of the year   **c.** $\frac{1}{4}$ of the year

4. For about what part of each year is there snow in Lakeside?

    a. $\frac{1}{6}$ of the year   **b.** $\frac{1}{4}$ of the year   **c.** $\frac{1}{2}$ of the year

5. Mountain View has sunshine for about what part of each year?

    a. $\frac{1}{4}$ of the year   **b.** $\frac{1}{3}$ of the year   **c.** $\frac{1}{2}$ of the year

6. Lakeside has rain for about what part of each year?

    a. $\frac{1}{2}$ of the year   **b.** $\frac{1}{4}$ of the year   **c.** $\frac{1}{3}$ of the year

7. Which city has more snow?

    a. Lakeside   **b.** Mountain View   **c.** about the same

8. Which city has more sunshine?

    a. Lakeside   **b.** Mountain View   **c.** about the same

B. Check your answers with another friend. Look at the problems that were difficult. Write a sentence that tells why.

Write Your Own Problems

A. Use the pictures to write problems about fractions. The problems can be about equivalent fractions, lowest-term fractions, comparing fractions, adding fractions, and subtracting fractions.

Problem 1

Equation(s): _____

Story: _____

_____

Question: _____

Problem 2

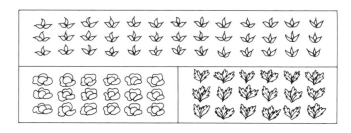

Equation(s): _____

Story: _____

_____

Question: _____

Problem 3

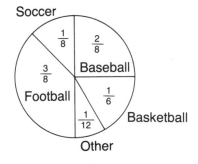

Equation(s): _____

Story: _____

_____

Question: _____

Problem 4

Equation(s): _____

Story: _____

_____

Question: _____

Problem 5

Equation(s): _____

Story: _____

_____

Question: _____

B. Work with two friends. Take turns reading and solving each other's problems. Use this space to solve your friends' problems.

C. Now check your answers with your friends. (You may use a calculator.) Look at the problems that were difficult. Write a sentence that tells what was difficult.

_____

Polygons

A. **Polygons** are plane figures with straight sides. **Vertices** are the points where the sides of a polygon meet. Study the names of these polygons:

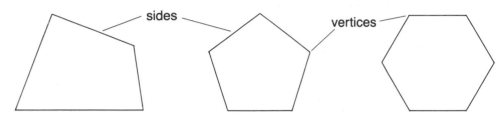 

| Quadrilateral | Pentagon | Hexagon | Octagon |

B. Work with a friend. Write a definition for each polygon.

1. A **quadrilateral** is _____

_____

2. A **pentagon** is _____

_____

3. A **hexagon** is _____

_____

4. An **octagon** is _____

_____

C. Work with a friend and answer the questions. (Write complete sentences.)

1. Is a square a polygon? _____

2. Is a circle a polygon? _____

3. How many vertices does a triangle have? _____

_____

4. How many vertices does a rectangle have? _____

_____

5. How many vertices does an octagon have? _____

_____

D. Sit with three or four friends. Talk about things at school and things at home that suggest plane figures. Write the names of the things at school in the left column and the things at home in the right column. Write the name of the plane figure in parentheses after each thing. The first two are done for you.

Plane Figures at School

clock (circle)

_____

_____

_____

_____

Plane Figures at Home

oven door (rectangle)

_____

_____

_____

_____

E. Work with a friend. You will need a ruler. Read the first five directions to your friend. Your friend will draw plane figures. Ask your friend to read the next five directions to you. Draw the correct plane figures.

You read to a friend:

1. Draw a triangle.

2. Draw a polygon with five sides.

3. Draw a figure that is not a polygon.

4. Draw a plane figure that has eight vertices.

5. Draw a quadrilateral.

Your friend reads to you:

1. Draw a figure that is made of a curved line.

2. Draw a pentagon.

3. Draw a figure with three vertices.

4. Draw a figure with six sides and six vertices.

5. Draw a figure that is not a polygon.

Drawings

Checkup

Complete the page by yourself.

A. In the middle column, write the fraction that is named. Then write each fraction next to the matching picture on the left and next to the matching equivalent fraction on the right.

_____    **1.** one-half _____   _____ $\frac{4}{12}$

_____    **2.** one-third _____   _____ $\frac{15}{20}$

_____    **3.** three-fourths _____   _____ $\frac{4}{6}$

_____    **4.** two-thirds _____   _____ $\frac{5}{10}$

B. Ring the correct phrase. Then write one of the signs (>, <, or =) in each circle.

1. one-half   is greater than / is less than / equals   one-third $\frac{1}{2}$ ◯ $\frac{1}{3}$

2. three-sixths   is greater than / is less than / equals   five-tenths $\frac{3}{6}$ ◯ $\frac{5}{10}$

3. two-thirds   is greater than / is less than / equals   five-sixths $\frac{2}{3}$ ◯ $\frac{5}{6}$

C. Solve each problem.

1. The school was $\frac{1}{2}$ mile from the Camera Shop and $\frac{3}{4}$ mile from the Ice Cream Shop. Which was farther from school? _____

2. Wei walked $\frac{1}{6}$ mile to Rob's house. Then they both walked $\frac{2}{3}$ mile to Dan's house. How far did Wei walk in all? _____

D. Write the name of each figure on the lines below. Choose from these names: **hexagon, quadrilateral, pentagon, octagon.**

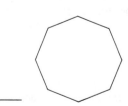

_____   _____

Learning Strategy Index

The lesson plans for the Student Book activities present and provide practice in the following learning strategies. For further information, please refer to the Teacher's Guide.

Metacognitive Strategies

Student Book pages

Advance Organization Previewing the main ideas and concepts of the material to be learned, often by skimming the text for the organizing principle.

5, 13, 15–16, 25, 31, 39, 53, 66, 67, 81

Organizational Planning Planning the parts, sequence, and main ideas to be expressed orally or in writing.

19–20, 29, 32–33, 45, 46–47, 48, 56, 57, 58, 60–61, 72–73, 74–75, 91–92

Selective Attention Attending to or scanning key words, numbers, phrases, linguistic markers, sentences, or types of information.

5, 7–8, 9, 10, 11, 12, 13, 14, 17, 18, 21, 22, 23, 25, 26–27, 28, 29, 30, 31, 32–33, 34, 38, 39, 40, 41, 42–43, 44, 46–47, 48, 50–51, 52, 53, 54, 56, 57, 60–61, 66, 67, 68, 69, 70–71, 74–75, 76, 77, 78, 79, 81, 82, 84, 85, 86, 87, 88, 89, 90, 95

Self-Monitoring Checking one's comprehension during listening or reading, or checking one's oral or written production while it is taking place.

6, 30, 44

Self-Evaluation Judging how well one has accomplished a learning task.

5, 6, 7–8, 9, 10, 11, 12, 13, 14, 15–16, 17, 18, 19–20, 21, 22, 23, 25, 26–27, 28, 29, 30, 31, 32–33, 34, 35–37, 38, 39, 40, 41, 42–43, 45, 46–47, 48, 49, 50–51, 52, 54, 55, 56, 57, 58, 59, 60–61, 62, 63, 64–65, 66, 68, 70–71, 72–73, 74–75, 76, 77, 78, 79, 82, 83, 84, 86, 87, 88, 89, 90, 91–92, 93–94, 95

Cognitive Strategies

Resourcing Using reference materials such as dictionaries, encyclopedias, and textbooks.

15–16, 17, 18, 19–20, 30, 32–33

Grouping Classifying words, terminology, numbers, or concepts according to their attributes.

5, 7–8, 9, 22, 53, 55, 56

Note-Taking Writing down key words and concepts in abbreviated verbal, graphic, or numerical form.

13, 19–20, 25, 30, 32–33, 39, 67, 81, 87

Summarizing Making a mental or written summary of information gained through listening or reading.

12, 22, 31, 38, 52, 57, 58, 59, 66, 72–73, 74–75, 79, 85, 86, 90, 93–94, 95

Imagery Using visual images (either mental or actual) to understand new information or to make a mental representation of a problem.

5, 9, 11, 14, 18, 21, 22, 29, 30, 32–33, 34, 35–37, 39, 44, 49, 50–51, 53, 55, 56, 59, 76, 77, 81, 82, 83, 84, 85, 86, 87, 88, 89, 90, 91–92, 93–94

Elaboration Relating new information to prior knowledge, relating different parts of new information to each other, or making meaningful personal associations to the new information.

5, 6, 10, 11, 12, 13, 15–16, 17, 19–20, 21, 22, 23, 25, 26–27, 28, 29, 30, 31, 32–33, 34, 35–37, 39, 40, 41, 42–43, 45, 46–47, 48, 49, 50–51, 52, 53, 54, 55, 56, 62, 63, 64–65, 66, 67, 68, 69, 76, 77, 78, 79, 81, 88, 89, 90, 93–94

Transfer Using what is already known about language to assist comprehension or production.

6, 7–8, 9, 10, 11, 13, 25, 49, 50–51, 54, 77

Inferencing Using information in the text to guess meanings, predict outcomes, or complete missing parts.

6, 11, 15–16, 28, 30, 31, 34, 40, 44, 53, 70–71, 90

Social Affective Strategies

Questioning for Clarification Eliciting from a teacher or peer additional explanation, rephrasing, examples, or verification.

6, 9, 12, 13, 14, 21, 22, 23, 25, 28, 29, 31, 34, 38, 39, 40, 41, 44, 52, 53, 54, 66, 69, 70–71, 74–75, 77, 79, 81, 82, 83, 84, 85, 87, 88, 89, 95

Cooperation Working together with peers to solve a problem, pool information, check a learning task, or get feedback on oral or written performance.

5, 6, 7–8, 9, 10, 11, 13, 14, 17, 18, 19–20, 21, 22, 25, 26–27, 28, 29, 30, 31, 32–33, 34, 39, 40, 41, 42–43, 44, 45, 46–47, 48, 49, 50–51, 53, 54, 55, 56, 57, 58, 59, 60–61, 62, 63, 64–65, 67, 68, 69, 70–71, 72–73, 74–75, 76, 77, 78, 81, 82, 83, 84, 85, 86, 87, 88, 89, 90, 91–92, 93–94

Self-Talk Reducing anxiety by using mental techniques that make one feel competent to do the learning task.

12, 23, 38, 52, 66, 79, 95